THE RISE OF EXPERIMENTAL BIOLOGY

THE RISE OF EXPERIMENTAL BIOLOGY

An Illustrated History

Peter L. Lutz, PhD
Florida Atlantic University, Boca Raton, FL

Foreword by
Bob Boutilier, PhD
Cambridge University, Cambridge, UK

HUMANA PRESS ✳ TOTOWA, NEW JERSEY

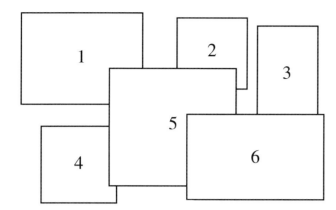

COVER PICTURE GUIDE AND PAGE REFERENCE

1. Auroch cave painting, p. 3
2. Right hand stencil, p. 3
3. Imhotep, Egyptian god of medicine, p. 12
4. Achilles binding the wound of Patroclus, p. 21
5. Late fifteenth century dissection, p. 70
6. Amputation of the leg without anesthetic, p. 157

© 2002 Humana Press Inc.
999 Riverview Drive, Suite 208
Totowa, New Jersey 07512

www.humanapress.com

This publication is printed on acid-free paper. ∞
ANSI Z39.48-1984 (American National Standards Institute)
Permanence of Paper for Printed Library Materials.

For additional copies, pricing for bulk purchases, and/or information about other Humana titles, contact Humana at the above address or at any of the following numbers: Tel.: 973-256-1699; Fax: 973-256-8341; E-mail: humana@humanapr.com or visit our Website: http://humanapress.com

Production Editor: Kim Hoather-Potter.

Cover Design by Patricia F. Cleary.

Printed in the United States of America.   10 9 8 7 6 5 4 3 2 1
Library of Congress Cataloging-in-Publication Data

Lutz, Peter L.
    The rise of experimental biology: an illustrated history/Peter L. Lutz
        p. cm.
    Includes bibliographical references and index.
    ISBN 0-89603-835-1 (alk. paper)
    1. Biology, Experimental – History. I. Title.

    QH324.L88 2002
    570'.7'24 – dc21

                                                              2001039830

FOREWORD

In this modern age of experimental biology and medicine, short-term perspectives overwhelm so much of our day-to-day endeavors that we often fail to acknowledge or indeed celebrate our rich and diverse history. With an engaging narrative that will have you turning "just one more page" well into the night, this book reveals to us the extent to which the modern scientific method has been shaped by the past. Though it is first and foremost a highly authoritative and scholarly account of the history of physiology and medicine, there are along the way many charming diversions where we are treated to some delightfully obscure anecdotes. The book is also hugely enriched by a treasure-trove of illustrations that chronicle a tortuous history of theoretical and technical developments ranging from the bizarre and amusing to the downright macabre. Peter Lutz takes us on a 30,000 year journey from the birth of the biological sciences, when we had already begun making graphic depictions of natural history on the walls of our caves, through to the foundations of contemporary physiology and medicine at the start of the twentieth century. He not only shows us how the biological sciences arose within the framework of the whole of the natural sciences, but reveals how our innate curiosity for rational explanations of life were often constrained by the social and political climate of the times. The journey has not always been a pretty one,

and Lutz sensitively tackles some of the most appalling and shameful episodes of human history to illustrate how advancements in biological science can and have been abused to satisfy repressive social or political imperatives. Such a history reminds us, particularly in this post-genomic era, of our obligation to society and to ourselves, to be vigilant guardians of our future. But what the book largely commemorates is the ancestral desire and continuing fascination we have to explain the living world around us. It reveals the trials and tribulations, the triumphs and disasters, the flashes of genius and the honest mistakes that have brought us to where we are today. It also serves in this day and age to remind us that it is curiosity-driven science that has had the greatest influence in founding our modern method, something we need to remember in an era of increasingly directed research.

This is a book that will entertain as well as inform a broad audience, from natural philosophers to experimental biologists to medical scientists. Indeed, it is written in such an accessible style and packaged in such a manageable size that it will make an excellent text book for any advanced undergraduate or graduate level course in the biological sciences. This is a celebration of the intellectual evolution of the science of physiology and medicine as well as being a socially conscious reminder of our privileged place in nature. I commend it to you.

Bob Boutilier, PhD
Department of Zoology
Cambridge University
Cambridge, UK

PREFACE

The impetus for this book comes from teaching animal physiology for many years. I found that though most life science and medical students were quite unaware of the antecedents of what they are being taught, they were interested and wanted to know more. *The Rise of Experimental Biology: An Illustrated History* attempts to satisfy this need by providing an overview of the origin and tortuous history of the science of biological function. An underlying theme is that science provides one of many explanations of the how and the why of life processes, but that it is unique because of its methodology—reason and experiment. However, there will be a running commentary that theories of biology do not develop in isolation, but are interrelated with their social context.

It is clearly impossible to hope to cover this story in a couple of hundred pages of text when hundreds of volumes by specialists have dealt with the details. Instead, I attempt to provide the merest sketches of this fascinating history. The elements are perhaps idiosyncratically selected: some conventional because as keystones they have to be mentioned, others more unusual because I find them amusing and particularly interesting. My approach then is not to give a conventional chronological list of who did what and when, but rather to provide a narrative on the highlights of the evolution of the science of biology.

An early, but important question I faced when organizing *The Rise of Experimental Biology: An Illustrated History* was how to handle references. In a conventional professional manuscript, each statement would be carefully annotated and cross-referenced. But in what is, one hopes, a popular book for general reading, such an arrangement would prove cumbersome; in any event, most of the details are common to the major science histories and readily available. Therefore, I have gathered all of my sources in the reference section of this book and list only particular references that give special information or peculiar points of view.

Peter Lutz, PhD
Department of Biological Sciences
Florida Atlantic University
Boca Raton, FL

Contents

List of Color Plates

Color plates 1–8 appear as an insert following p. 82.

Plate 1 (Fig. 1.1 from Chapter 1). Graphic descriptions of nature are the most ancient human intellectual activity known. Illustrations from the Chauvet caves in Southern France, which contain the oldest known cave paintings, about 30, 000 years old (Chauvet et al., 1996). Reproduced with permission of the French Ministry of Culture and Communication, Regional District for Cultural Affairs-Rhône-Alpes Region, Regional Department of Archeology. (A) Fine renderings of heads of the long-extinct auroch. (B) A herd of rhinoceroses. The decreasing size of the horns and the multiplication of the lines depicting the backs suggest perspective. (C) Stencil of the right hand of a 30,000-year-old individual reveal a concept of self. The outline was made by pulverizing pigment on the hand flattened against the wall.

Plate 2 (Fig. 1.4 from Chapter 1). Trephination is still practiced today in parts of Africa, e.g., by the Kisii tribe in Kenya. (Reproduced with permission from Mueller and Fitch, 1994.) (A) A cross-shaped incision through the skin is made with a razor, the scalp flaps are reflected, and a hole is scraped through the skull bone with a hack saw. (B) After the surgery, the skin flaps are replaced and smeared with petroleum jelly.

Plate 3 (Fig. 9.5 from Chapter 9). A late fifteenth century dissection. The bleeding indicates that the corpse is fresh. The inevitable accompanying dog looks on. (From *De propietatibus rerum* by Bartholomeus, Bibliothéque National, Paris.)

Plate 4 (Fig. 16.1 from Chapter 16). Hunter's transplant of a cock's spur into a cock's comb. (From Royal Society of Surgeons, England.)

Plate 5 (Fig. 16.4 from Chapter 16). Anatomical display from the Hunterian Museum, Glasgow University (S. Milton).

Plate 6 (Fig. 16.5 from Chapter 16). Mortsafes: Early nineteenth century graves armored against "resurrectionists." St. Mungo's Cathedral, Glasgow.

Plate 7 (Fig. 18.1 from Chapter 18). Louis Agassiz's "physiological" interpretation of God's plan on the unfolding of life on earth for the preparation of man. (From Agassiz and Gould, 1851.)

Plate 8 (Fig. 19.5 from Chapter 19). Eugenics Society poster warning about the dangers of broadcasting bad seed.

1   Introduction

Science Is Improbable

The Rise of Experimental Biology is meant to serve as a brief introduction to the rich and sometimes amusing history of experimental biology. Today, compared to only 50 years ago, we have learned so much about how the body works that we are on the verge of an unprecedented ability to intervene in body function and disease, an advance that rivals any of the great human historical achievements. Some are afraid that we have gone too far, and that the latest advances in molecular biology now threaten to challenge our core beliefs about human life and worth. In a recent edition of the journal *Nature,* for example, Peter Aldhous (2000) discusses the possibility of "therapeutic cloning," in which a cloned identical-twin embryo could be grown from a patient's own healthy cell, to serve as a source of perfectly matched stem cells for future organ replacement or repair—an idea quickly and roundly condemned by Pope John Paul II and other religious authorities, as well as many ethicists and scientists. How did we get here? The answer is fascinating.

Although the earliest cultural artifacts speak to a committed human interest in nature that dates back at least 30,000 years, our current knowledge results from an explosion of experimentally generated information that started only 300–400 years ago. To appreciate what happened, why it happened, and to understand the birth and maturation of the biological sciences, requires some idea of its history. Unfortunately, biology is commonly taught as if modern ideas spring into existence without precedence and without any historical perspective. The student is unaware of the background of trials and struggles, brilliant insights, well-meaning mistakes, or entrenched conservatism and dogmatism that are all factors in building and developing testable and rational explanations for Life. Priestly's gross misinterpretation of his groundbreaking experiments—showing that plants produced oxygen and animals consumed oxygen—is a classic case in point (*see* Chapter 15).

Without an appreciation of how the experimental method produced a revolutionary and powerful means to advance knowledge, scientific explanations of natural phenomena might be no more convincing than the accounts provided by ancient tales of mythology and mysticism. Indeed some contemporary post-modernist philosophers argue that all explanations represent opinions, and that all opinions are equally valid (*see* Chapter 20). Any history of science that is not merely a catalog of events will show that the problems scientists choose to work on can be strongly influenced by contemporary societal interests and priorities. As we will explore, however, the history of the growth and maturation of the scientific method is intrinsically different from histories of, for example, battles, fashions, or fashionable ideas. As science progresses we obtain more and more accurate information about the composition of the universe, living and non living, and deeper insights into its workings.

Throughout the course of *The Rise of Experimental Biology,* we will continually discuss what science actually is. For now, we can take the workmanlike definition of Crowther (1938), who wrote that science is "the system of behavior by which man acquires mastery of his environment." This is accomplished by providing increasingly more accurate explanations of natural phenomena. In recent years we have been compelled to add to this concept Bernal's assertion (1965) that science, however free, must be funded. The paymasters want useful knowledge, and the bottom line is that the more accurate the knowledge, the greater its usefulness. Finally, we can round out our definition with the view that science is not just practicality or technology, but importantly, something that enriches "our lives by providing a more powerful view of ourselves and our world" (Vogel, 1992), an intellectual role that motivates many scientists.

We might well see the rudiments of scientific interest in the "mastery over the environment" or the manipulative intervention in natural processes, motivations that largely define the human condition. For instance, data from sites in China and Africa show that humans made domestic fires 240,000 years ago. Potassium-argon dating of fire-

places discovered in Kenya suggest that the use of fire might even go back as far as 1.6 million years (Rowlett, 1999). It has been suggested, on such evidence, that Homo erectus had the ability to cook otherwise indigestible tubers by making fires, which led to an important expansion in the diet base of early humans and thus played a key role in human evolution.

Graphic depictions of nature are the most ancient human intellectual activity of which we know. The most recently discovered (1994) cave paintings in the Chaveux region of France are the oldest known in the world, dated by radiocarbon assay from a painting of a rhinoceros to be about 32,000 years old (Chauvet et al., 1996). The Chaveux cave paintings are as sophisticated as those in other illustrated prehistoric caves, such as the famous Lascaux caves (discovered in 1940), but twice their age, and it is reasonable to suppose that they are based on an even more ancient tradition. The Chaveux cave galleries depict with astonishing vigor and skill a diverse menagerie of animals. There are over 300 paintings of black charcoal and red ochre, as well as scored-out rock engravings. They depict hunted and hunting animals, including rhinoceroses, horses, lions, bears, owls, and mammoths (Fig 1.1A). According to Chauvet et al. (1996), some of the paintings resemble each other so closely that the same individual must have created them. As is common to cave paintings throughout the world, abstract geometric patterns are also depicted (Fig 1.1B), and in a very personal touch, clusters and individual red palm prints of 30,000 year old men, women, and children are seen throughout the cave (Fig 1.1C) (Balter, 1999).

The purpose or function of Paleolithic cave and rock art has long been a source of controversy, but the fact that it was practiced for over 30,000 years argues that it served some basic needs. Food gatherers require practical knowledge of their environment-when and where prey will occur, how to recognize spoors, how to trap, and kill, how to appease the animal spirit world, and how to protect oneself from harm. This information must be learned and transmitted and it has been suggested

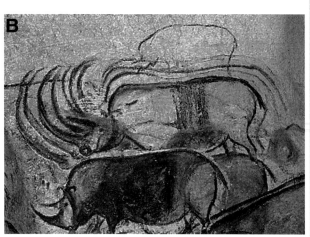

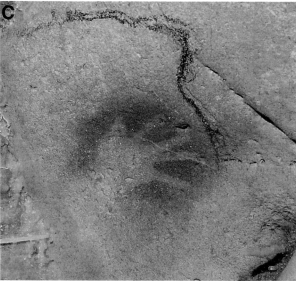

Fig. 1.1. Graphic descriptions of nature are the most ancient human intellectual activity known. Illustrations from the Chauvet caves in Southern France, which contain the oldest known cave paintings, about 30,000 years old (Chauvet et al., 1996). Reproduced with permission of the French Ministry of Culture and Communication, Regional District for Cultural Affairs-Rhône-Alpes Region, Regional Department of Archeology. (**A**) Fine renderings of heads of the long-extinct auroch. (**B**) A herd of rhinoceroses. The decreasing size of the horns and the multiplication of the lines depicting the backs suggest perspective. (**C**) Stencil of the right hand of a 30,000-year-old individual reveal a concept of self. The outline was made by pulverizing pigment on the hand flattened against the wall. (*See* color plates appearing in the insert following p. 82.)

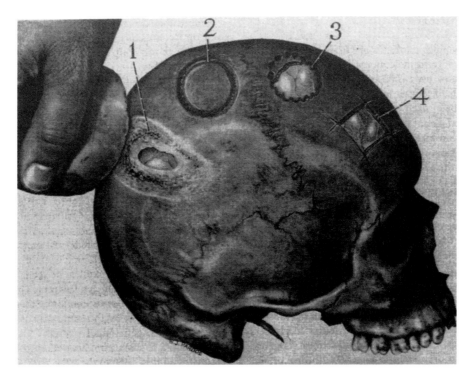

Fig. 1.2. Boring a hole in the skull, trephination, appears to be an attempt to treat (mental) sickness, and is by far the most ancient medical surgery known. Prehistoric methods of skull trephiny (**1**) scraping, (**2**) grooving, (**3**) boring, (**4**) rectangular incisions (From Rose, 1997). Reproduced with permission by © Swets and Zeitlinger.

that cave/rock paintings served just such an educational function. More particularly, archaeologist Jean Clottes argues that some of the paintings function to provide a window into the animal spirit world (Chauvet et al., 1996). The abstract dots and zigzag patterns, a common and universal feature of rock art, may relate to the tribal shaman's hallucinogenic entrance into the animal world depicted on the cave walls (Bower, 1996). Native American, African, and other modern-day shamans often report that in the first stage of trance they perceive certain basic geometric forms such as grids, dots, zigzags, and concentric circles (Bower, 1996). They then enter into a state of hallucinogenic visions in which they are transformed into animals, to be one with nature and to receive the wisdom of the animal spirit world for the benefit of the tribe. Human-animal motifs are common to cave and rock art throughout the world at all ages. Even the Chauvet caves show a half-bison, half-man figure.

Cave paintings indicate a deep interest, or even a compulsion, to depict nature, but they provide no evidence of any attempt to understand how nature actually works. A distinguishing characteristic of the earliest science, or protoscience, is the attempt to intervene in a natural process through some type of understanding of how it works. Such intervention requires technique, and precisely in technique is the origin of science: Its source is experience, its aims are practical, and its only test is what works. From this socially functional stand point, practical knowledge-science-develops in a close correspondence with the stages of human social progress (Bernal, 1965).

Some of the earliest evidence of surgical intervention in humans may be seen in skulls, which are the most frequently found and best-preserved part of the human body recovered from archaeological excavations. Prehistoric skulls with deliberately made holes are found in graves throughout the world, indicating the widespread use of an early

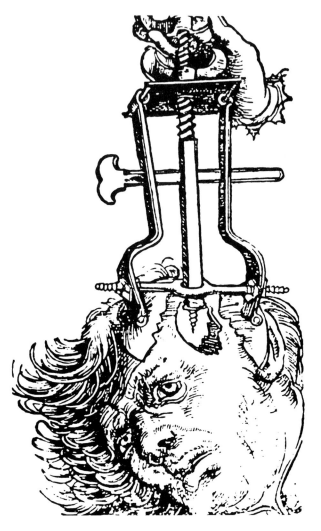

Fig. 1.3. Trephiny was still a major remedy in the sixteenth century. Heroic trephiny in the Renaissance (1528).

surgical technique called trephination. This practice appears to have been an attempt to treat mental sickness, and is by far the most ancient medical surgery known. Unlike ritual mutilation, such as facial scarring or circumcision, which are performed on healthy people for religious or cultural reasons, trephination was an attempt to heal a manifest illness. Hardly a trivial procedure, it involved working without asepsis or anesthetics, slicing open the highly vascularized skin covering the skull, scraping or boring the bone to make a hole-typically about two centimeters or more wide-and then closing the wound (Fig 1.2). The cure was thought to be effected by releasing evil spirits, or, in modern terms, by relieving subdural pressure.

One might think that such deeply invasive surgery would result in high mortality, which would have been the expected outcome in the civilized West up until about 100 years ago. Yet many ancient trephined sculls bear evidence of new-growth healing, showing that the patients must have enjoyed long-term postoperative survival.

Trephined skulls are found throughout the world, on all continents. The oldest are found in Europe, dating from at least 10,000 to perhaps 35,000–100,000 years ago (Rose, 1997). The oldest in the Americas, about 8000 years old, come from southern Peru and Northern Chile, where the practice seems to have been relatively common. Trephination has had an astonishing endurance. It

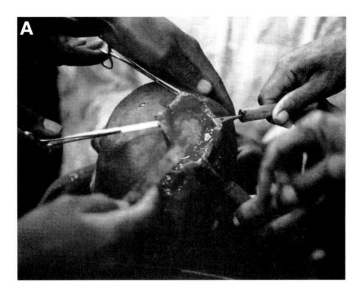

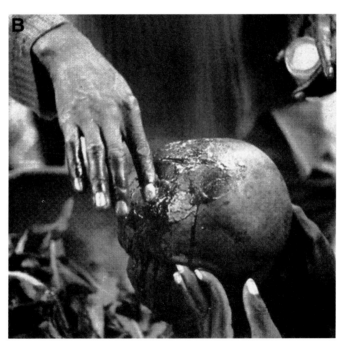

Fig. 1.4. Trephination is still practiced today in parts of Africa, e.g., by the Kisii tribe in Kenya. (Reproduced with permission from Mueller and Fitch, 1994.) **(A)** A cross-shaped incision through the skin is made with a razor, the scalp flaps are reflected, and a hole is scraped through the skull bone with a hack saw. **(B)** After the surgery, the skin flaps are replaced and smeared with petroleum jelly. (*See* color plates appearing in the insert following p. 82.)

was practiced in ancient Greece; for example, around 400 BC Hippocrates recommended the surgery for specific head wounds (contusions, fissure fractures, and indentations). As late as the sixteenth century, the famous French physician Ambroise Paré (1510–1590) described trephination in detail as an important surgical remedy (Fig 1.3).

Trephination is still practiced today in parts of Africa (Rawlings and Rossitch 1994), and until recently 500–800 trephinations per year were performed in the Kisii tribe of Kenya, mainly for acute cranial trauma (violent head injuries) and posttraumatic headache. Muller and Fitch (1994) provided a fascinating description of the contemporary Kisii version this ancient surgical technique. A cross-shaped incision is made with a razor through the skin covering the scull down to the bone; the scalp flaps are reflected, and a hole is scraped in the skull bone with a hacksaw (Fig 1.4A). After surgery, the flaps are replaced and smeared with petroleum jelly (Fig 1.4B)

Such attempts to heal through physical intervention presume a theory of how the illness was caused. The antiquity and universality of trephination indicate that an interest in the causes of body malfunction goes back to our most ancient common roots.

But science had a religious and spiritual function that was not at all trivial. In fact, as we shall see the primary societal justification for science throughout most of its history has been that it illuminated humankind's relationship with the universe and revealed details of God's Great Design. Because it had such an exalted role, science was called the "handmaiden" of religion in the Middle Ages. As J. James points out in his excellent book, *The Music of the Spheres* (1993), until very recently the universe was seen as an intellectually comfort-able, comprehensible, anthropocentric system in which everything made sense. The Greeks had painstakingly developed the concept that all that we see, hear, and know is an aspect of the ultimate truth and has the simplicity of a geometric theorem. Order ruled over the Earth and the heavens above. The heavenly bodies revolved in sublime harmony, producing an exalted but real music of the spheres, the music of the universe. For the religious the predictability of the movements of heavenly bodies was thought to be a reflection of the essential perfection of the universe, and the diversity of the living world revealed the majesty of God's Glorious Design. Scientists, philosophers and artists had a common higher purpose: To reveal the details of this universal harmony. Up until the eighteenth century, many—if not most—scientists justified their studies on religious grounds, believing that their purpose was to illuminate God's wonderful creation. It was not until the nineteenth century that scientists abandoned such noble ends, and began to be satisfied with more mundane, immediately functional, purposes and explanations.

2 The Beginnings

It is part of the human condition to attempt to describe, explain, and control nature. This is seen in the Paleolithic representations of animals and ancient skull surgery. In the Mesopotamian civilization, more accurate anatomical descriptions serve more sophisticated divinational purposes. In early ancient Egypt, a specialized system of medical practice is supported by more detailed anatomy and speculation on the function of some body parts.

MYTH AND MAGIC

The compulsion to describe nature, for whatever reasons, the need to interfere physically to alleviate illness, and the drive to control one's destiny are deeply engrained human activities with roots at the beginnings of homo sapiens: The evidence is universal and ancient. For the vast majority of humankind, throughout almost all of history, natural events, such as the rising and setting of the sun and the waxing and waning of the moon, are compelling illustrations of vital significance. They are signs or omens of humans' place in an all-encompassing drama of life, a drama that involves both natural and supernatural forces.

Explanations of natural phenomena are immediate and complete in themselves. Take, for example, the terrifying phenomenon of lightning: Who did it? In the myths of Northern Europe, the answer is, the god, Thor. Why did he do it? To show great anger, as witnessed in the flash and the roar. How did he do it? By throwing a magic thunderbolt, or by striking his gigantic mallet on a great flintstone. If any further explanation is needed, Thor is a member of a large family of gods who fit into an all-embracing assemblage of supernatural beings, which gives similar explanations of natural events. These divine beings provide an overarching cosmic structure that is responsible for the past, determines the present, and controls the future. The

9

Fig. 2.1. Religious explanations of natural phenomena such as thunder are universal. Wooden sculpture of the Nigerian Yoruba tribe god of thunder, Shango, holding his spear. (From Peter Lutz, 2001.)

mere description of the phenomenon suffices; no further details are needed. The importance of the phenomenon lies in its relevance to the supernatural scheme of nature, and this is what is explained and argued about. Such explanations are global and ancient, and clearly satisfy a basic human need. The thunder god is universal and tenacious: The original Zeus of Greece and Guamansuri of ancient Peru; in Nigeria, the traditional god of war and thunder of the Yoruba tribe is Shango, who carries a lightning spear (Fig. 2.1). This god is still wor-

shipped today in the African-derived cults of Central America and Brazil, and in the present-day Santeria religion of Cuba. His likeness can been purchased today in any Santeria (Botanica) shop in the Little Havana district of Miami. In English-speaking countries, we still commemorate Thor in Thursday.

But besides providing interpretations of natural phenomena, the religious belief system had to work, in the sense that it had to contain kernels of practical knowledge for supplying essential needs

(Bernal, 1965). Although diverse cultures talked a great deal of nonsense about explanations, those that survived acted with a great deal of sense. This is a necessary condition, i.e., the group's actions with respect to getting food, providing protection from disease, making tools, weapons, fortifications, and so on—all must be practical and successful to the degree that allows the tribe, city, empire to survive against natural forces and other human rivals. They must be able to predict when critically important seasonal changes will happen, such as when winter starts, when rain will come, or when to gather and when to sow the crop. To this end, the priest has sacred formulas to encourage or compel the gods to cause the sun to rise each day, or to end winter. Through this secret knowledge the priest has the power to change the future by invoking prayers that will avert, defeat or deflect pestilence. These functional requirements are embedded in all of the widely diverse particulars of the mythological formulas that regulate law and behavior in different cultures throughout the world. However, early civilizations share one common feature: The most important and essential rituals are practiced in specialized powerhouses, the temples, by expert professionals, the priests.

As societies became more complex, so did their religions. In sophisticated ancient Rome, it was believed that the world was full of little gods that had to be invoked on special occasions, even to the minutiae of life. There was the god, Vatican, who caused the infant to utter his first cry; Fabulinus, who prompted his first word; and Cuba, who guarded him when old enough to exchange the cradle for cot (Osler, 1921). The priests were responsible for interpreting exquisitely detailed omens. Specialized Roman augurs read the will of the gods from the patterns of flying birds and their sounds. But even more, there was one set of priests for eagles and hawks, and another set for swifts and owls. No war was undertaken and no decision made, no matter how small, until the priests had read the omens.

In contrast to religious dogma, natural explanations of the natural world, its properties, and their changes, do not depend on, or even use, the intervention of a god or spirit. In contrast to mythologically driven accounts, a close examination of the phenomenon is important: Natural explanations are dependent on detail. Initially, the most plausible explanation is the one that is in closest agreement with current knowledge of natural forces. This is established by argument. Later, the proposed explanations are subjected to increasingly rigorous tests of demonstration and experimentation. Here we are involved in a chain of explanations, in which the immediate natural cause has a natural cause, which itself must have a natural cause, and so on. In principle, we are on a path *ad infinitum*, or one that halts at some ultimate cause.

However, building up a corpus of description and explanation of the natural world requires the luxury of a group of people that has the leisure not to work for a living, so that they can spend their time reading, writing, and discussing: a philosopher class. The freedom of this privileged group depended on the stable social organization of the city-state, the material wealth of which was supported by the labor of an underclass, usually bound slaves. The development of a system of record-keeping, writing, which could transmit knowledge from generation to generation, was a critical element in this process. Such conditions arose in the ancient civilizations of China, the Indus Valley, Mesopotamia, and Egypt. Once a critical population mass was reached, technical revolutions had to provide the essential material bases of these civilizations. An increased population density could only be supported by higher productivity. Consequent practical innovations included the domestication of animals, higher-yield agriculture through the selection of grain crops, and the techniques of pottery, brick making, and metal work. For example, through the experience of trial and error, the Egyptian smiths gradually developed a high practical knowledge of metallurgy. They found that the hardest bronze was made of 12% tin and 88% copper, that decreasing the tin content resulted in a softer bronze, and that increasing it resulted in a more fragile product (Bernal, 1965). Increasingly accurate anatomical descriptions also appear in these early civilizations, but they mainly serve in divination.

MESOPOTAMIA

Although civilization flourished in ancient India and China, the strongest roots for the future development of science lie in the Middle East, the true start of our story.

The area between the Tigris and Euphrates rivers, in what is now Iraq, is the site of ancient Mesopotamia, homeland of one of the world's first civilizations. Settlements there have been dated as far back as 10,000 BC. About 5000 BC, the Sumerian civilization was established. One of the greatest rulers of Mesopotamia, Hammurabi, who ruled from his capital Babylon about 1790–1750 BC, caused a massive pillar of stone to be erected, which contained the Codex of Hammurabi. Although primarily a legal document, concerned with property rights, it also laid down the law for medical and veterinary practice. A sliding fee schedule is given of payments and penalties for treatments to gentlemen, slaves, and domestic animals (Sarton, 1952). The Codex provides a fascinating picture of the robust dangers to both doctor and patient of ancient surgery, and describes the first managed health care: "If the doctor shall treat a gentleman and shall open an abscess with a bronze lancet and shall preserve the eye of the patient, he shall receive ten shekels of silver. If the doctor shall open an abscess with a blunt knife and shall kill the patient or shall destroy the sight of the eye, his hands shall be cut off or his eye be put out" (Gardner, 1965).

Veterinary surgeons were similarly regulated, but less vulnerable to such drastic malpractice consequences: "If a veterinary surgeon perform a major operation on an ox or an ass and has caused its death, he shall give the owner of the ox or the ass one-fourth of its value" (Sarton, 1952).

Several hundred cuneiform tablets concerned with medical issues have survived. These deal mostly with prescriptions for illnesses, but a few describe diseases, the causes of which were typically spirits (one spirit for one type of ailment), and

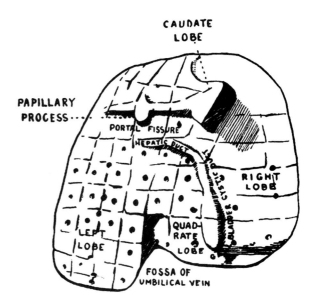

Fig. 2.2. By as early as 3000 BC, hepatoscopy had become an extraordinarily complex art in Babylonian prognostication. A clay model of a sheep's liver used to teach divination in ancient Babylonia, about 2000 BC. Anatomical features are labeled by Singer (1957). Reproduced with permission from Dover Publications.

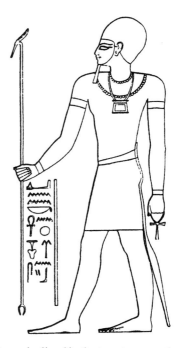

Fig. 2.3. Imhotep, who lived in the twenty-seventh century BC, was later worshipped as the god of medicine in Egypt, and identified with the Greek god of medicine, Asclepius. A representation of Imhotep, the Egyptian god of medicine, Temple of Ptah at Karnak. (From Hurry, 1928.)

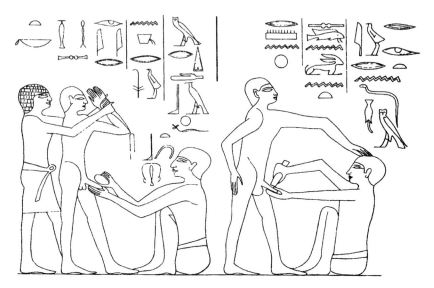

Fig. 2.4. The earliest representation of circumcision. From the Necropolis of Sakkara, dated at the beginning of sixth dynasty. (From Castiglioni, 1958.) A stone knife was used for the operation.

the treatments often aimed at the expulsion of the malingering spirits (Frey, 1975).

There is abundant evidence of an interest in anatomy in the Babylonian-Sumerian Empire. But the details are sharply pragmatic—almost all are connected to the art of divination, which is the telling of the future through inspection of the organs of sacrificed animals.

Organ divination was an important and specialized skill, one that required many years of training. Of all the organs, the liver, from its size, central position, and richness in blood, was the most impressive. Furthermore, for generations blood was believed to be the essence of life, so the blood-rich liver was naturally regarded as the seat of life. Although other organs of the sacrificial animal had their own specialized uses, reading the liver—hepatoscopy—was the most important way to learn the divine will.

By as early as 3000 BC, hepatoscopy had become an extraordinarily complex art in Babylonian prognostication. The lobes, gallbladder, swellings of the upper lobe, and their markings, were all carefully inspected by the priest. A large number of texts were written on this topic, and many illustrative and anatomically accurate clay models sur-

vive, made to assist or teach liver divination. The earliest known, dating from about 2000 BC, is a model of a sheep's liver, with an included divination text (Fig. 2.2). In many of these models, the liver is divided into lobes, then into squares, with each square having its own special prognostication. Surviving texts provide some of the rules that the priests used in their interpretations. For example, a gallbladder, swollen on the right side, indicated superiority in the king's army, and was favorable; but swelling on the left side indicated success of the enemy, and was unfavorable. A long bile duct pointed to a long life (Jastrow, 1935).

EGYPT

Ancient Egypt is the first civilization with written records that show an extensive interest in medicine. However, like all of the early civilizations, Egyptian biological science was essentially empirical and pragmatic, dealing directly with magic and healing, and with little interest in theory.

A hierarchical medical profession was established early in Egypt. Imhotep (Fig. 2.3), who lived in the twenty-seventh century BC, was considered by Osler (1921) as "the first figure of a physician to stand out clearly from the mists of antiquity." He

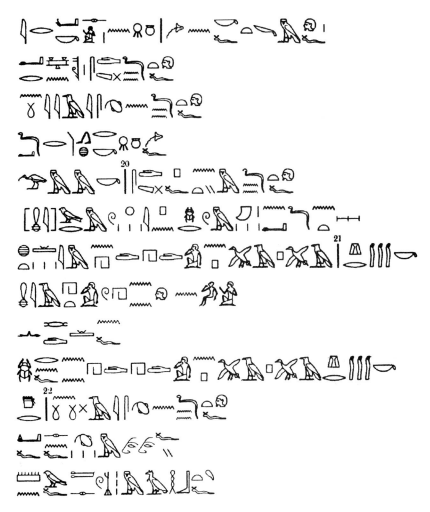

Fig. 2.5. Hieroglyphic of case six in the Edwin Smith surgical papyrus. (From Breasted, 1930.)

was chief minister to Djoser, the second king of Egypt's third dynasty. In later times, Imhotep was worshipped as the god of medicine in Egypt, and was identified with the Greek god of medicine, Asclepius, in Greece. A physician class soon developed that became powerful and specialized. There were physicians of the eye, stomach-bowel physicians, and "guardians of the anus" (Sarton, 1952).

The Egyptian concept of life differed from that of Mesopotamian medicine in that while the latter saw the liver as the source of life-giving blood, the Egyptians apparently gave a greater emphasis to the act of respiration as central to life. They believed, quite accurately, that life failed when respiration stopped.

But blood was still important. *The Book of the Dead* relates how the gods Hu and Lia arose from the blood that gushed from the sun god Ra, when he cut off his penis, and mummies were painted red to give them the strength of blood (Castiglioni, 1958).

Many medical papyri have been preserved, indicating the vast amount of biological/medical knowledge in ancient Egypt. The oldest are a half-dozen or so from the period between 2000 and 1500 BC, which are probably based on older texts dating to around 3000 BC (Frey, 1975). One of the oldest, the Kahun medical papyrus (ca. 2000 BC), is the most ancient document on gynecology known. The largest, the Ebers papyrus, is 20 m

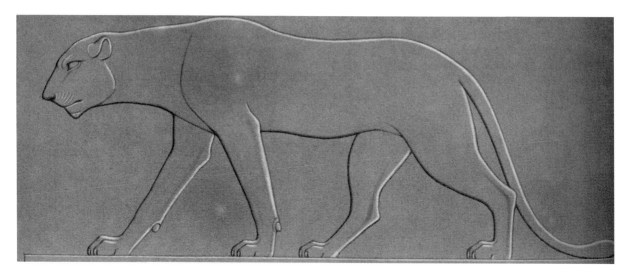

Fig. 2.6. Some early Egyptian animal sculptures show a beautiful naturalism. A 3500 year old bas-relief, from the temple of Assassif, of an Egyptian panther.

long, with more than 2000 lines. It begins grandly: "Here begins the book of the preparations of medicines for all parts of the bodies of a person" (Stevens, 1966). The Ebers papyrus contains 877 recipes or prescriptions for a great variety of diseases or symptoms, systematically arranged and treated in an orderly fashion, and some instructions for surgery. For example, according to the Ebers papyrus, male circumcision was performed at the age of fourteen (Fig. 2.4). The circumcision of girls was also of general use in ancient Egypt (Castiglioni, 1958).

The most famous medical papyrus, the Edwin Smith surgical papyrus, written about 1700 BC, is much shorter, being only about 5 m long, and is available to us in a widely quoted translation by Breasted (1930). It deals, not with general recipes, but with definite cases. The Smith papyrus contains a systematic treatment of 48 cases of injury and wounds, starting at the head and going to the shoulder region. It is likely that the rest of the body was dealt with in a continuation of the manuscript, now lost (Sarton, 1952).

The cases are systematically discussed under the following formal arrangement: title, examination, diagnosis, treatment, and glosses (an explanation of technical terms used). In 13 cases, the opinion is that the wound is fatal, and not to be treated, i.e., there is a scientific interest in recording and discussing observable facts for which there is no treatment (translation from Breasted, 1930).

One example (case six [Fig. 2.5]) deals with a gaping wound on the head, penetrating the bone and revealing the brain.

Title: Instructions concerning a gaping wound in his head, penetrating to the bone, smashing his skull, (and) rendering open the brain of his skull.

Examination: If you examine a man having a gaping wound in his head, penetrating to the bone, smashing his skull, and rendering open the brain of his skull, you should palpitate his wound. Should you find that smash which is in his skull like those corrugations which form in molten copper, and something therein throbbing, fluttering under your fingers, like the weak place of an infant's crown before it becomes whole when it has happened there is no throbbing and fluttering under thy fingers until the brain of (the patient) is rent open-and he discharges blood from both nostrils, and he suffers with stiffness in the neck.

Diagnosis: following this detailed description is rather an anti-climax: "an ailment not to be treated."

Egyptian medicine incorporated many fundamental physiological observations about the living body that had been made from ancient times. But the Egyptians' very complicated religion ruled all. For them, the distinction between life and death was of especially deep concern. The living body is warm, breathes, and moves. Death is the absence of these qualities, and the corpse is cold and still. The heart was the center of life. It is the source of life-giving heat and the house of the soul. The rapid beating of the heart in anger, fear, and surprise indicated that these emotions arose in the heart. It was believed that the pulse was the heart "speaking" to the rest of body through the blood vessels (French, 1978a). The Egyptians, accordingly, treated the heart with special reverence during embalming: It was elaborately wrapped and carefully preserved in a special jar, or placed back in the body cavity. The brain, by contrast, was considered worthless. It was removed part by part by a hook through the nose, and discarded.

Some early Egyptian animal sculptures show a beautiful naturalism (Fig. 2.6). In the period of the Middle Kingdom (starting about 2000 BC), however, the nascent rational and empirical spirit seen in the Edwin Smith papyrus gradually waned and Egyptian science became fossilized, subsumed into magic and mysticism. An example of mystical logic, which will be referred to later; the number 4 became especially revered; there were four columns to support the roof of the temple, four sides of the pyramid, and medicine was to be taken four times a day. We see then that in some early civilizations, such as in Mesopotamia and Egypt, there was an initial attempt to ascribe natural causes to some natural phenomena. But the effort was not sustainable and was engulfed by overwhelming, conservative religious forces.

3 Early Greek Science

CONTENTS

Fresh scientific concepts begin with the Greeks, and they are of astonishing originality. For the first time in history there was a sustained attempt to supply purely natural interpretations of the universe as a whole, living and material. Ingenious natural schemes flourished, which rivaled the established mythologies, giving instead, nonreligious explanations of cosmology, the origins of life, and the functions of the human body. The investigators called themselves "physiologia," observers of nature, or "natural philosophers." A few examples illustrate the rich imaginativeness of their earliest speculations.

THE MILESIANS

About the sixth century BC, early Greek intellectual activity centered on the small independent coastal towns of the Adriatic basin. The cities were run by a mercantile aristocracy, and had a comparatively simple political structure. For merchants, logical and pragmatic problem-solvers were in favor, and the rich new city-states tolerated sophisticated debates about the nature of society, arguments over the best form of government, and speculations on the nature of the universe. A rich variety of imaginative explanations flourished about how the universe was structured, but, uniquely, the Ionian philosophers attempted to explain all natural phenomena in terms of rational reasoning, rather than relying upon the gods.

The earliest and most substantial records are associated with the ancient Greek city of Miletus. This port city, situated in western Anatolia, had more than 60 trading colonies throughout the Mediterranean and Black Sea, and a mercantile fleet and land caravans that brought both material and intellectual goods from as far as India and China. The imaginations of the early Milesian philosophers flourished under such favorable sociopolitical circumstances.

Thales (ca. 624–548 BC), the first of the pre-Socratic philosophers, reputedly had received some

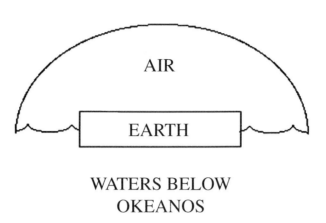

Fig. 3.1. Thales' conception of the floating earth. He envisioned Earth as a flat disk floating on water, with a bowl or firmament of water above our heads.

education in Egypt. His most famous concept was that all was originally water (the *arche*), and that land was formed from this primordial water by a natural process, like the silting up of the Nile Delta. In his scheme, earth, air, and living things separated in the beginning, by a process similar to that described in the Sumerian creation myth and the book of *Genesis*. But Thales left out the creator. He envisioned Earth as a flat disk floating on water, with a bowl or firmament of water above our heads, where rain comes from (Fig. 3.1). Thales explained that earthquakes were not caused by the direct intervention of gods, but were the result of earth rocking in the water on which it floats, like a boat in a storm. This was a rather courageous natural explanation of what had always been considered a terrifying act of divine retribution (Farrington, 1944).

Expanding on his theme, Thales regarded water as the life-giver. He drew attention, for example, to the moist natures of life-sustaining food and life-giving semen. Yet water itself was not sufficient for life: Living animals had to have a soul, because the soul was the cause of motion in things. However, this soul was an inherent property of matter, not a quality given from outside (Hall, 1969a).

A contemporary, and perhaps pupil, of Thales, Anaximander (ca. 610–545 BC), developed a more complex scheme of cosmology. Anaximander conceived the idea of the First as an indefinite (boundless) substance called *aperion*, from which the four basic elements (fire, air, earth, and water) were derived. He proposed that, in the beginning, these elements stratified by weight. Earth, the heaviest, was at the center, covered by a shell of water. This shell in turn was blanketed by mist (for a very long time, air, mist, and vapor were thought to be the same substance), and finally an envelope of fire surrounded all. However, once this arrangement was established, the natural properties of the elements produced change. Fire heated the water, *causing* evaporation, and resulting in the appearance of dry land. The increase in the volume of mist *caused* increasing pressure, which *caused* the fiery envelope of the universe to burst and take the form of wheels of fire. These wheels are enclosed in tubes of mist circling around the earth and sky. This scheme provided a natural explanation for the stars and their motion. The heavenly bodies we see circling above our heads are holes in the tubes, which move in different directions and at different speeds (Farrington, 1944). The importance of this explanation is that it is an ingenious *working model* of the universe, with interconnected causes and effects.

Anaximander also gave an evolutionary explanation for life and its origin: Animals arose from moisture that was undergoing evaporation through the action of fire. The first living forms, such as fish, lived in water, but, when land appeared, some fish moved to land and became land animals. Even humans developed from watery ancestors. The first humans spent their infant lives as aquatic creatures (like the fetus still does), and, when they came out on land, they were covered with a sort of bark (perhaps like fish skin) to protect themselves. However, later, as the bark dried, cracks appeared through which men and women were delivered (Hall, 1969).

Of the writings of Anaximenes (ca. 550 BC), the last of the great Milesians, only three fragments survive, but they were very influential. Anaximenes was not impressed by the fire idea; he countered that everything is permeated by air (or mist), and everything is ultimately air. Air is soul, air is divine; even the gods arose out of it. In the natural world, air exists according to its state of rarefaction and condensation. The most rarefied air is fire; condensed air is water, and, as it becomes more condensed, the water turns into earth. Anaximenes declared that rarefaction is always accompanied by heat, condensation by cold. And, in a new way of thinking, he provided an ingenious *experimental* demonstration of his hypothesis: Breathe gently on your hand, and you can feel the warmth caused by the rarefied vapor; blow hard, and the condensed vapor produces cold.

THE POST-MILESIANS

The intellectual breakthrough of the Milesians in seeking natural explanations for the universe inspired philosophers throughout Greece. A mystical, but nonreligious theoretical approach to nature was provided by Pythagoras (ca. 582 BC). About 530 BC, he settled in the Greek colony of Croton in southern Italy, and founded his brotherhood or sect. The Pythagoreans maintained that truth could be revealed by deductive reasoning, and they developed an elaborate philosophy based on the concept that number is the principle of all things. The sort of reasoning involved was highly sophisticated for the

time, and became extremely influential. For example, for the Pythagoreans, like the Egyptians, the number four had a special significance. Four was the

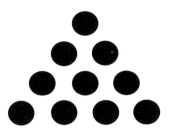

smallest number of points needed to construct the sides of a pyramid, a perfect solid.

Adding these points together gives 10, the perfect Pythagorean number. Such numerology was used to interpret the universe. For example, since there must be perfection in the universe, the number 10 must rule there. Philolaus, a follower of the Pythagorean school, considered the sun to be at the center of the universe, and was the first to propose the motion of the earth around the sun. Reckoning the number of heavenly bodies in universe, he counted one sun and eight revolving bodies. The revolving bodies consisted of: one sphere of fixed stars (1), five planets (5), the moon (1) and Earth (1). But, since this only adds up to nine universal bodies, and, from Pythagoras we know there must be 10, there must be another body that we do not see. Philolaus's solution is a counterearth revolving around the sun, which we can never observe (Lloyd, 1970). This ingenious hypothesis has all the elements of a self-consistent theory that can be falsified—but the idea of testing was to wait almost 2000 years.

The musical universe was another seminal Pythagorean concept. To Pythagoras, music was number (proportion), the universe was number, so the universe was music. He distinguished between three sorts of music. The first kind, ordinary music, was made by instruments. But there is a music of the body that we cannot hear, musica humana, made by

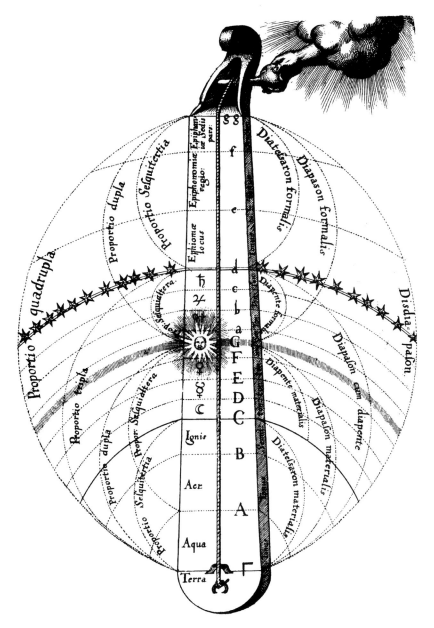

Fig. 3.2. To Pythagoras, music was number (proportion), the universe was number, so the universe was music. The concept is illustrated in Thomas Fludd's (1547–1637) *Musical Harmony of the Universe* (from *Monochordium mundi* [1623]). A Pythagorean monochord is divided into the basic harmonic intervals, each of which is an element of the universe. The earth is low G.

the resonance between the soul and the body. Disharmony in musica humana produced agitated mental states; harmony was conducive to tranquility. And, finally, the greatest of all music, musica mundana, the music of the spheres, was made by the cosmos itself, by the heavenly bodies revolving at distances proportionate to a musical (Pythagorean) scale (James, 1993). According to Aristotle, the reason none of us can hear this celestial music is because "the sound is in our ears from the very moment of birth and is thus indistinguishable from its contrary silence, since sound and silence are discriminated by mutual contrast." Aristotle's analogy was that of the coppersmith gradually growing oblivious to the clanging of his own hammer (James, 1993). The idea of the universe constructed as a heavenly harmony had great poetic force, and lasted well into the seventeenth century (Fig. 3.2).

Some Pythagoreans also undertook biological investigations. Alcemon of Croton (ca. 450 BC), who wrote a treatise, *Peri Physeos*, has been called the first neuroscientist (Gross, 1999). He appears to have dissected the brain, since he noted that the optic nerves came together. He may have performed vivisection, since he believed that the brain was the site of sensation and cognition. More particularly, he thought that there were some pathways between the brain and the sense organs, which if broken, stopped communication (French, 1978a).

The writings of Heraclitus (ca. 500 BC) are only preserved in obscure fragments, but they were very influential. Heraclitus emphasized change or flux in nature. He argued that fire was the first principle, that all arises out of fire, and all is ultimately reconverted to fire. More imaginatively, he developed the doctrine of opposite tensions; he said that there is a force that moves all material objects up toward fire, and there is an opposite force that moves them down to earth. The state of matter in any particular configuration is the result of the balance between these two opposing forces of tension. Because matter is not stable, and change is the primary feature of the cosmos, the world is continually involved in the transformations of fire into water and water into earth. All is in flux, all is changing. His most famous attributed aphorism (as described in Plato's *Theaetetus* [Russell, 1945]) is, "You cannot step into the same river twice." According to Heraclitus, life is a balance or harmony between the continuing tension of the opposites of living and dying. Although man has a special place in this scheme, because he has a soul, he is nevertheless part of the world, because the soul comes from the same heavenly fire (Hall, 1969a).

Empedocles (ca. 504–432 BC) provided an extremely influential account of living nature. He was a leading member of the medical center at

Fig. 3.3. A fifth century BC illustration of Achilles binding the wound of Patroclus. (From the National Library of Medicine.)

Agrigentum, an important center of Greek medicine. Empedocles taught that nature consists of mixtures of four elements or roots: fire, air, earth, and water, and that the elements are composed of combinations of four qualities, hot, cold, wet, and dry. In this system, water = cold + wet, fire = hot + dry, earth = cold + dry, and air = hot + wet.

The transmutation of an element could occur by altering the proportions of its qualities: For example, heating water will drive out cold, turning it into air. More abstractly, he held that the four elements are influenced by the metaphysically opposing entities of Love and Strife: when Love dominates, the elements are in harmony, but Strife produces discord. Humans fitted into this scheme, because humans and nature are governed by the same rules.

Empedocles believed that the blood was the medium of thought, and that the degree of a person's intelligence depended on the composition of his blood (Gross, 1999). This blood was contained in fleshy tubes that open out at fine terminations in the skin, and which draw in and expel air as the blood ebbs and flows. He believed that semen was a form of blood, and that sex was determined by the quality of the semen (Sullivan, 1996a). Empedocles was also interested in natural explanations of how the senses worked. For example, it had long been noted that, when the eyeball is pressed under conditions of total darkness, flashes of light are seen, now called "deformation phosphenes." He included this observation in his theory of vision. He used the fact of deformation phosphenes, and the observation that the eyes of some animals, like cats, light up in the dark, to support his suggestion that light is generated in the eye. Visual perception, he argued, is caused by this light shining out and being reflected back to the eye by the object (Grüsser and Hagner, 1990). The belief that vision depended on light emanating from the eye survived in various forms until the Renaissance (Fig. 3.3).

Democritus (460–370 BC) modified Empedocles' theory, to suggest that visual perception was caused by the interaction of internal fire generated by the eye and an external fire emanating from the regarded object. More famously, Democritus taught that everything in the universe is made up of atoms of different shapes. The soul (*psyche*) is made of the lightest atoms, which are especially numerous in the brain, which is thus the seat of the mind. Cruder atoms are concentrated in the heart, the center of emotion, and the most coarse atoms are in the liver, the seat of lust and appetite. In various guises, this trichotomy in organs and functions had a long influence on physiological thought.

For Diogenes of Apollonia (ca. 440–430 BC), the brain is the center of life. Air, the fundamental substance of the universe, is drawn into the brain through the nose, and leaves its best parts there, bestowing sense perception and understanding. It then passes to the rest of the body with the blood in the vessels.

An effective account (in the sense of influence and endurance) of the human condition was provided by Hippocrates (ca. 460–370 BC) and the Hippocratic school of medicine, as recorded in the texts of the *Corpus Hippocraticum.* Although the Greeks had a long tradition of practical knowl-

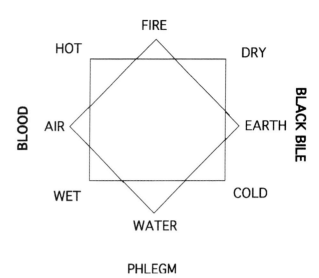

Fig. 3.4. A diagrammatic representation of the Hippocratic balance of the four humors and elements (based on Singer, 1957). The Empedoclean four-element system was enriched by the addition of four biological humors: blood, yellow bile, black bile, and mucus (phlegm). These four humors relate directly to the four elements, and the humors were responsible for the four basic human temperaments: sanguine, choleric, melancholic, and phlegmatic.

edge of the healing arts (Fig. 3.3), a basic theoretical assumption of the Hippocratic school is that the human body is governed by the same rules as the natural world. The Empedoclean four-element system was enriched by the addition of four biological humors: blood, yellow bile, black bile, and mucus (phlegm). These four humors relate directly to the four elements, and the humors were responsible for the four basic human temperaments: sanguine, choleric, melancholic, and phlegmatic (Fig. 3.4).

The mixture of the four humors in an individual determined personality and state of health. When the humors were in perfect balance (*krasis*, or harmony) all was well; when out of balance (*dyskrasis*, or distemper) illness ruled. This concept allowed the construction of a straightforward, self-consistent system of apparently rational medicine, the aim of which was to restore the natural balance.

Since illness results when the humors are out of balance, the task of medicine is to identify the precise cause of the imbalance, and to restore it.

The Hippocratic treatise, *On Regimen* has elaborate recipes to this end. For example, too much fire produces an increased choleric disposition, and this causes an excessive quickness of perception, and can lead ultimately to madness. The remedy is to restore the fire–water balance. For a half-mad person, a restorative diet includes eating boiled vegetables and drinking only water, or, rather kindly, if water is not tolerated, soft white wine will suffice.

By having simple explanations for all illnesses, and simple, "logical" treatments, the humoral theory proved popular among physicians, and became a dominant influence in Western medicine and physiology for the next 2000 years. Just how "rational" or revolutionary the Hippocratic approach could be is illustrated in the book on the "sacred disease," or the falling sickness, epilepsy. Epilepsy was universally regarded as a powerful manifestation of possession by evil spirits, but, in the Hippocratic treatise titled *Sacred Diseases* it is stated that the falling sickness "is in my opinion no more divine or sacred than any other diseases but has a natural cause." Fitting in with the humoral theory, it is suggested that the disease comes from a stoppage of air reaching the brain, through an excess of phlegm (Sarton, 1952).

And, finally, in the Hippocratic system, the brain had a special place. It was the cognitive and sensory organ, the seat of pleasure, merriment, and amusements, as well as grief, pain, anxiety, and tears. It is the organ that enables us to think, see, and hear (Gross, 1999).

THE SOUL

The soul had a special place in Greek science. The Greeks believed that the ultimate properties of life lay in a vital principle they called the "soul." As far back as Homeric times, it was thought that there were at least two distinct souls in the body, the pneuma (or *thymos*) and the psyche (or *nous*). The pneuma, or breath soul, was responsible for vital, life-giving heat and movement, and dwelt in the chest (French, 1978a). But this breath/air soul had a much grander role: It permeated the universe, and was responsible for the life of all organisms. It was breathed in with the first breath of the infant, and was released back to the world soul with the last breath of the dying human. In *Sacred Diseases* it is mentioned that, when a man draws breath (*pneuma*) into himself, the air (*aer*) first reaches the brain, from where it is dispersed throughout the rest of the body.

Only humans had the highest soul, the psyche, which was the self, the "me", with the qualities of consciousness and thought. It was not clear if the psyche resided in the head or the heart of the living body, but it was immortal. This early distinction between the different souls and their different locations was later greatly elaborated. For example, the philosopher, Philolaus, had the mind (*nous*) housed in the brain, and the soul (*psyche*) rested in the heart.

SUMMARY

The Greeks were the first to attempt to explain the world in rational and concrete, rather than religious terms. In a novel way, "understanding was to be its own reward" (Wolpert, 2000). Perhaps the most important contribution that the early Greeks made to the development of science was that their competing rational explanations allowed, for the first time, rational criticism. The Greeks believed that truth would be discovered by reasoned argument. For example, Thales held that first was water, from which a seed of hot and cold separated, forming the cosmos, which grows like a living being. But, if the first was water, how could its opposite, fire, come into existence? It was debated that, if, as Anaximander argued, water holds the earth up, then what holds the water up? One solution was that Earth hangs free, remaining where it is, because of its equal distance from everything; the universe is boundless, and there is no reason why Earth should fall in any direction. For Heraclitus, all is in a state of flux, but Parmonides argued that change does not occur, and Democritus made all of unchanging atoms.

For the Greeks, internal consistency of an argument was the critical factor, which would lead directly to drawing *necessary* inferences from given premises. Indeed, if the properly premised proposition produces a conclusion that appears contrary to experience (such as the nonexistence of movement), then we must ignore our experience and have confidence in reason (*logos*), not allow the eye, ear, and tongue to rule. This approach eventually led to the belief that truth could only be obtained through logical argument. The economy and consistency of the argument are critical; empirical data obtained through the senses were secondary, and might not be not trustworthy.

4   The Age of Plato and Aristotle

In covering the classic Greek contribution to science we could enumerate many steps; for example, Diocles of Carystus (ca. 350 BC) founded the Dogmatic School of Medicine, and around 300 BC, wrote the first known anatomy book, based on animal dissection and observation of human subjects. But our aim is to select some seminal concepts that influenced future developments in biology. From this perspective, Aristotle and Plato are by far the most important of the Greek philosophers: They dominate physiological thought for the next thousand years, and are the sources of enduring inspiration and long-lasting misconceptions.

PLATO AND THEORY

Plato (429–347 BC) a contemporary of Hippocrates, founded his school, the Academy, where symposia were held. The original symposia must have been lively affairs, alcohol-stimulated, free-wheeling arguments on philosophical propositions, such as "the world that we know through our senses is an illusion."

His treatise *Timaeus*, subtitled a "likely tale," presents an organization of the universe and the human body that incorporates both mythical and scientific elements in a rational account. Geometrical construction, proportionality, and ratios rule. For example, he states that equilateral triangles are the common elements of fire, air, and water, and that the earth is composed of squares. But Plato's main interest in science was to reveal the operation of reason in the universe (Lloyd, 1970). Biological discussions are scarce, and are purely speculative, and the language is often figurative and mystical (Sullivan, 1996b). Sometimes, by chance, they were correct, and are forgotten. For example, he placed the seat of thought and feeling in the brain—but his reason was that this duty befitted the organ nearest the heavens. Sometimes they were incorrect, and endure. For example, he considered the role of producing new life as an exalted office, appropriate to the brain, and not the low testes. Seed-bearing semen was therefore made in the

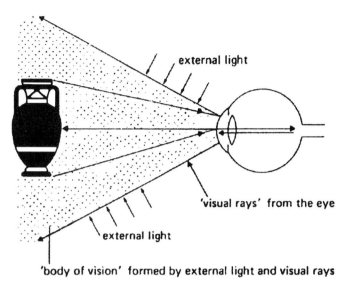

Fig. 4.1. Plato's theory of vision. Visual rays are emitted from the eye and interact with external light to form a cone of vision. When this cone touches an object, rays are reflected back to the eye. (From Grüsser and Hagner, 1990.)

brain and transmitted to the penis through an open passage in the base of the vertebra. As late as the sixteenth century Leonardo da Vinci drew two passages to the penis, one from the testes and one from the spinal marrow (Fig. 10.2).

Yet Plato was much more interested in the absolute than the concrete, and the crux of the Platonist approach was the concept of the true world versus the sensible world. Knowledge of the true and eternal world of ideal forms is only attainable by gods and a few enlightened philosophers. Knowledge obtained through the senses is imperfect or illusory, because sense perception is fraught with indeterminable error. Like Empedocles, Plato held that vision is the result of light issuing from the eye as from a lantern. This light interacts with external light to form a "cone of vision," which on touching an object, is reflected back to the eye (Fig. 4.1; Grüsser and Hagner, 1990). But we have no way of knowing how much this light is distorted. In *Phaedro*, he said that if we are to know anything absolutely, we must be free from the body, and behold actual realities with the eye of the soul alone. In the *Republic*, he declared that the starry heavens are to be apprehended by reason and intel-

ligence, not by sight. He taught that we see only the superficial appearances, reflections or shadows of perfect forms (the archetypes) of objects.

Plato speculates on the structure of living forms. The soul comes in three parts: reason, the highest part, lives in the brain, where it "lords over the all the rest"; the mortal soul, responsible for life, lives in the heart; and the lowest soul, appetite is placed in the liver, "tethered like a beast."

Plato had strong opinions on sex and sexual appetite. The problem was the uncontrollable, willful, sexual organs, which appeared to have minds of their own. They are sometimes "disobedient and self willed." The uterus, in particular, roamed about women's bodies, leaping from place to place. "When it remains unfertilized for a long time the uterus becomes very angry and moves all over the body, by obstructing the outlets of the pneuma it prevents respiration, and throws the woman in confusion" (*Timaeus*) (Translations from Jowett, 1953). The Greek word for womb is "hysterikos," from which the English word "hysteria" is descended.

Plato drew a parallel between the construction of the inner world of the human body, later called the "microcosm", and that of the outer world, the "macrocosm," a scheme, as we will see, that had great endurance. But in the end, as perhaps Plato would have agreed, his scientific explanations are ultimately mere deceptions, at best "likely tales," a position, as we shall see later, that has recently blossomed anew in some departments of philosophy. Plato's dismissal of the truth value of empirical science was to dominate western philosophy until the twelfth century, when Aristotle's teachings started to filter back from the Islamic world.

ARISTOTLE AND OBSERVATION

Aristotle (384–322 BC) a pupil of Plato, and tutor to Alexander the great, founded his school, the Lyceum, in 335 BC. In contrast to Plato, Aristotle had a strong interest in describing and explaining living nature. His approach was empirical, and he was especially interested in learning by enquiry or research (*historiá*). He emphasized the importance of appealing to facts in sufficient numbers, before forming general conclusions. This of course meant finding the facts. Most unusually, his interest in nature was predominantly intellectual, and he was not particularly interested in medicine or cures.

As a philosopher, Aristotle theorized on the nature of scientific knowledge. He asserted that there is an intrinsic importance and value in the study of nature, using reason (*nous*), contemplation (*theoria*), and observation. The object of a science is to find principles, elements, and causes, and the proper procedure is to go from complex effects to simple causes. Through the use of deductive argument, knowledge demonstrates connections that are necessary, eternal, and universal.

Aristotle proposed four types of causation, or factors, involved in things that exist in the material world: the material cause, or what the body is made of, its components; the formal cause, or the body's shape, the configuration of its components; the efficient cause, or the cause of change such as movement or growth; and the final cause, or the purpose of the body: Note that this is a natural cause, not divine.

Aristotle was preeminently a teleologist: In much of his biological writings, his aim was to prove the presence of a purpose in nature. Nature, he believed, is not just a collection of chance happenings; animals have purposes, and every body organ serves some partial end toward that purpose. He had no sympathy with evolutionary explanations that called on chance or accident to account for animal characteristics. He believed that Empedocles was in error when he said many of the characteristics of animals are merely the result of accidental occurrences during development. Using the analogy of building, Aristotle argued that, because the true object of architecture is not the bricks or mortar or timber, but the house, so, with biology, every body organ serves some partial end for the body as a whole. Everything in nature knows its place, and for the most part keeps to it. It is the nature of a bird to fly in air, or of a fish to swim in water.

Aristotle accounts for natural motion through the concept of natural place. If an element is removed from its natural place it tends to return to that place i.e., stones and rain fall, air and fire rise. But he was not immune to abstract deduction. He argued, for example, that Empedocles's four elements are not sufficient to account for the circular motion of the heavens. All complex motion is a combination of three elementary types of motion: upward, downward, and circular. Upward and downward motion can be derived from a combination of the four elements, but not circular. Reason, therefore, tells him that there must be a fifth element, the quintessence (*quinta essentia* from Latin) or ether (from Greek), the translucent sphere of the cosmos, an indestructible eternal whose natural motion is circular.

Especially in his later works, Aristotle showed a strong interest in biology, displaying an astonishing richness in observation and speculation on the world of nature. Over one-fifth of his extant work concerns biological topics.* One can gather from

*Translations are from Ross (1942) unless otherwise stated.

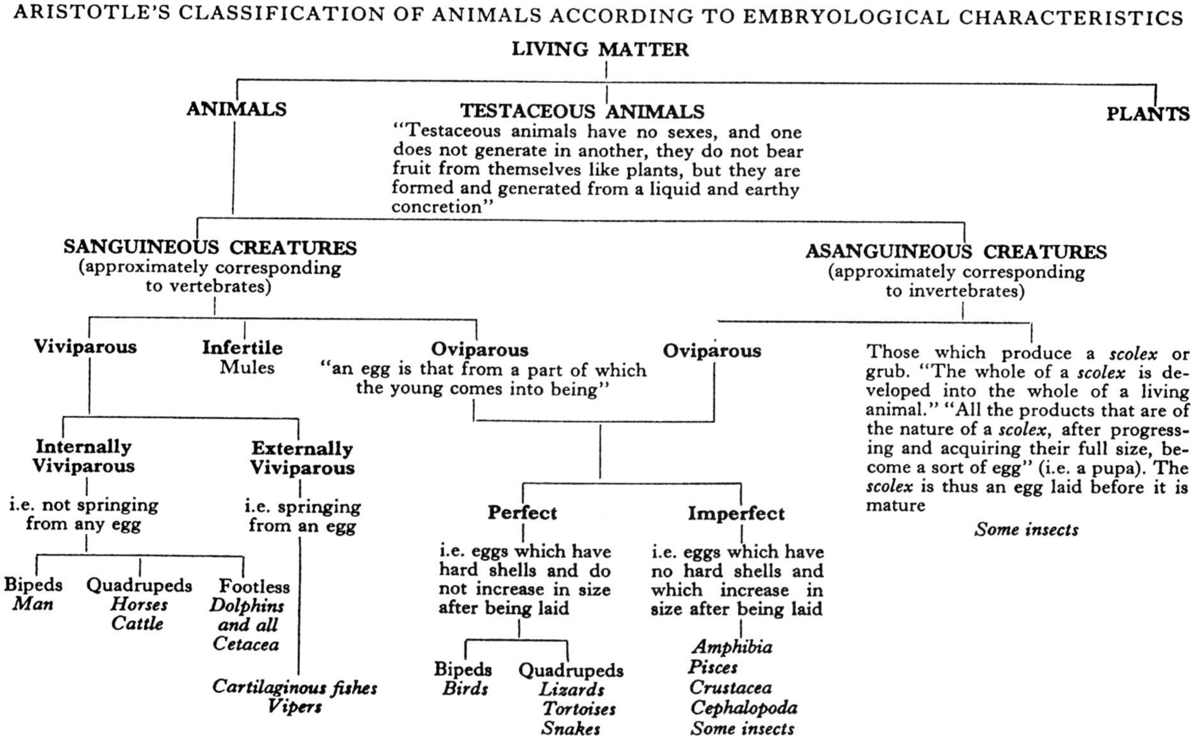

ARISTOTLE'S CLASSIFICATION OF ANIMALS ACCORDING TO EMBRYOLOGICAL CHARACTERISTICS

LIVING MATTER

ANIMALS   **TESTACEOUS ANIMALS**   **PLANTS**
"Testaceous animals have no sexes, and one
does not generate in another, they do not bear
fruit from themselves like plants, but they are
formed and generated from a liquid and earthy
concretion"

SANGUINEOUS CREATURES   **ASANGUINEOUS CREATURES**
(approximately corresponding   (approximately corresponding
to vertebrates)   to invertebrates)

Viviparous   **Infertile**   **Oviparous**   **Oviparous**   Those which produce a *scolex* or
                 Mules   "an egg is that from a part of which   grub. "The whole of a *scolex* is de-
                               the young comes into being"   veloped into the whole of a living
                                                                                        animal." "All the products that are of
                                                                                        the nature of a *scolex*, after progress-
Internally   **Externally**   ing and acquiring their full size, be-
Viviparous   **Viviparous**   come a sort of egg" (i.e. a pupa). The
i.e. not springing   i.e. springing   **Perfect**   **Imperfect**   *scolex* is thus an egg laid before it is
from any egg   from an egg   mature
                                     i.e. eggs which have   i.e. eggs which have   *Some insects*
                                     hard shells and do   no hard shells and
Bipeds Quadrupeds Footless   not increase in size   which increase in
Man   *Horses*   *Dolphins*   after being laid   size after being laid
        Cattle   *and all*
                    Cetacea   *Amphibia*
                                     Bipeds   Quadrupeds   *Pisces*
        Cartilaginous fishes   *Birds* *Lizards*   *Crustacea*
                Vipers   *Tortoises*   *Cephalopoda*
                                              Snakes   *Some insects*

Fig. 4.2. Aristotle's classification of animals according to the possession of (red) blood and embryological development. (From Needham, 1950.) Reproduced with permission from Cambridge University Press.

comments in the texts that some of these books originally had illustrations, all long lost.

The book on the *Generation of Animals* (*De Generatione Animalium*) deals with reproduction; *Movement of Animals* has a detailed discussion of modes of locomotion; and on *Respiration* (*De Respiratione*) concerns the mechanics and purpose of breathing. In the *History of Animals* (*Historica Animalium*) he gave detailed descriptions of appearances, habits, and characteristics of more than 500 animals, including 120 kinds of fish and 60 kinds of insects.

Aristotle found a natural pattern to the bewildering variety of animals. This was revealed when animals were divided into red-blooded and nonred-blooded types, which corresponds approximately to vertebrates and invertebrates, and were classified according to their embryological characteristics, which produced novel affinities, such as that between birds and reptiles (Fig. 4.2). Species

(*eidos*) were distinguished and gathered into higher groupings (*genus*), in a system that essentially survived until the early eighteenth century. Man is included in this scheme, with the rather humble distinguishing feature of being bipedal.

The book, *On the Parts of Animals (De Partibus Animalium)*, is an anatomical study of the animal as composed of its parts, in which an attempt is made to assign a particular function to each part of the body. The books *Of Youth and Old Age,* and *On Life and Death (De Juventute et senectute et de Vita et Morte)* discuss animal heat and the function of respiration.

Aristotle states that knowledge of external parts is not enough; that we need to dissect to "recognize how a thing is constructed or has come to be;" that we need to learn by active enquiry. His justification was truly extraordinary: This activity of finding out how things are "provides extraordinary pleasure ..., it belongs to what is beautiful," (Lloyd, 1970) and he declared that, if any person thinks the examina-

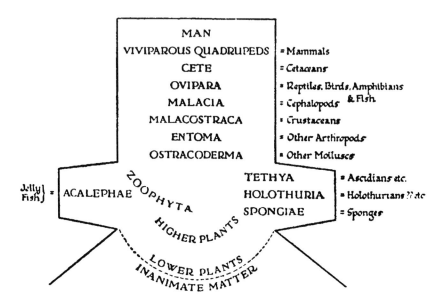

Fig. 4.3. A representation of Aristotle's *scala natura*. (From Singer, 1957.) Nature is envisioned as an orderly ladder with progressive steps, from the inanimate to plants, from the lower animals to the higher, to man and spirits, and finally God. Man is half animal, half spirit. Reproduced with permission from Dover Publications.

tion of the rest of the animal kingdom an unworthy task, he must hold in scorn the study of man.

He was especially interested in marine animals, and described the structure and habits of sponges, ascidians, holothurians, starfish, crustaceans, and cephalopods. He distinguished cetaceans from fish, and differentiated between live-bearing and egg-bearing reproduction. His description of viviparity in the dogfish, which he claimed that the fetus had a placenta-like attachment, was dismissed as error, until rediscovered by Muller in 1842 (Singer, 1957).

He discussed whether human intelligence is the consequence or cause of possessing hands (and concludes the latter, that hands are instruments to serve the purpose of intelligence). Some observations certainly would have required dissection, especially the description of the structure of the eye of a mole and the stomach of ruminants.

Despite the great diversity of his observations, Aristotle, as a philosopher could not resist seeing an ordered pattern in nature, the great chain of being, or the "Scala Natura." Nature was envisioned as an orderly ladder with progressive steps

from the inanimate to plants, from the lower animals to the higher, to man and spirits, and finally God. Humans are half animal, half spirit (Fig. 4.3).

Aristotle described, discussed, and theorized body function in extraordinary detail, but was always influenced by "rational" argument. For example, he held that the kidneys were not necessary for life, because they were not seen in all animals, such as fishes and birds.

Aristotle considered the three souls as the active forces in his physiology: the vegetative soul for nourishment and growth, the animal soul for motion and sensation, and the rational soul for intellect (Hall, 1969a).

Heart, Brain, and Blood

Aristotle agreed with Empedocles that the heart was the central organ, the seat of perception, intelligence, and vitality, the source of the nutritive blood, and the site of animal heat. His justifications for the heart's preeminence are reasonable, and are based in part on observation. The heart is the first organ to appear in the developing embryo, "the heart beats like a speck of blood in the white of the egg ... The heart is the first structure to be endowed with motion ... It orchestrates further differentiation and organization of the growing body" (Frampton, 1991). Aristotle argued that the source of innate heat must be the heart, because, when the warmth of the heart is quenched (compared to other organs), death ensues. Also, the heart's central location "naturally" makes it best situated to be the source of nutrient-giving blood to the rest of the body.

Aristotle was scathing toward those who proposed that the brain was important for sensation, pointing out that the brain is devoid of all feeling, is bloodless, and is naturally cold to the touch (hinting at experimentation). Moreover, the function of the brain could be arrived at by the use of reason (French, 1978a). Since "every body member serves some partial end"... and as all influences require to be counterbalanced so that they may be reduced to moderation and brought to the mean, nature has contrived the brain as a counterpoise to the heart with its contained heat," and "the brain tempers the heat and seething of the heart" (*On the Parts of Animals*). Or, being the moistest and coldest part of the body, the brain had the duty of cooling the heart "the hottest of all bodily parts." However, it had one other important function: In chilling the blood, the brain causes sleep. He further pointed out that the heart was affected by emotions, the brain not at all. All animals have sensations, but only vertebrates and cephalopods have brains. The heart is first to form and last to stop working. And man's brain is the largest and moistest, because man's heart is the hottest. Since the heart has supreme control, Aristotle admonishes that physicians should concentrate on the heart, and not on the subordinate organs.

At that time, it was commonly thought, and Aristotle agreed, that the blood was contained in the heart and blood vessels, as in a vase; hence, the use of the term "vessel." The heart's thick walls and hollow cavity made it ideal for holding the blood (its task). The walls of the heart, likened to the walls of a furnace, also serve to protect the vital heat that is formed in the heart. The pulsation of the heart is "similar to boiling," caused by this innate heat expanding the blood. However, the innate heat must be moderated, or else it will burst into self-consuming flames. The necessary cooling is accomplished by air from the lungs, carried to the heart by hollow air tubes, the pulmonary vessels.

For a long time, the arteries (air vessels) were believed to be hollow air-holding tubes. The arteries were thought to be extensions of the trachea, another air tube, which was known until the seventeenth century as the "arteria aspera" or "trachea arteria," (the rough artery, from its cartilaginous texture). It was believed that the pulsing in the arteries was caused by the movement of this air. This mistake, that the arteries hold air, probably came from observations on hanged cattle that had been slaughtered by having their throats cut: Blood would have drained out of the major arteries, but still be held in the veins.

The body therefore had two types of vessel, each with a distinctive function. The purpose of the venous vessels was to distribute the nutritive blood to the body; the function of the arterial vessels was to transmit the respiratory spirit to the body and cooling air to the heart. The heat of the hot blood was then cooled by the brain.

Assimilation

Aristotle was especially interested in how the nutritional quality of food was transmitted to blood, and how this nutrient was then changed into tissue. He asked, "What agency governs the assimilative process?" At one level, the answer was thought to lie in the nature of the nutritive soul, the psyche. The simile used by Aristotle is water or grape juice changing or "growing" into wine. The nutritive soul is the agent that converts food (potential flesh) into actual flesh. But it is not sufficient just to say this change happens; it is necessary to describe exactly how it occurs. A systematic process is proposed. The purpose of the mouth is to soften and break food into small pieces, in order to more easily allow a "concoction" (*pepsis*), or a separation of the food into its component parts, to take place in the stomach. This word is still in use today, as "pepsin," a stomach enzyme that aids in protein digestion. The first pepsis is brought about through the action of an internal moist heat (Boylan, 1982), rather like the cooking of food or the ripening of fruit. The products are passed to the intestine, where the concoction is conveyed to the liver and spleen via the mesentery blood vessels (being sucked up, like the roots of a plant draw up water). The next pepsis occurs in the liver and its companion organ, the spleen, which Aristotle calls a counterbalancing, bastard liver. Here, with heat as the agent, the food is changed into primary blood. Liquid wastes from the liver, spleen, and blood vessels are sent to the kidneys and bladder, although the liver has a special waste product, the bile.

The heart receives the twice-concocted blood food, and, under the action of the heart's intense heat, together with the addition of the animal soul, the pneuma, the final nutriment is formed. This higher blood exits the heart, via the venous blood vessels, to the organs, by oozing through the blood vessels "like water in unbaked pottery," or the blood moves like a stream in an irrigation ditch. In this one-way process, the blood is used up in the organs as it is transubstantiated into tissue. A last pepsis takes place in the testes, where semen is made (Boylan, 1982).

Respiration

Aristotle's *De Respiratione* is one of his most fascinating biological works, in which he demonstrates a wide-ranging knowledge of comparative biology and a prolific intellectual curiosity about life. The basic premise is that the function of respiration is to cool or moderate animal heat (French, 1978a): "Every animal in order to prevent itself from dying requires refrigeration." Given that the heart is the source of this warmth, which, when quenched, results in death, then, in order to conserve body heat, there must be some way of cooling the heart. The analogy given in *On Youth and Old Age* is that of a charcoal brazier. With the choker on, and the air supply diminished, the fire is quickly extinguished; but, if the brazier is cooled by lifting the lid up and down quickly, the heat is conserved, and the coals remain glowing for a long time. The refrigerating operation of respiration is witnessed when one breathes in cold air and breathes out warm (the only known difference between inspired and expired air, therefore, it must have a cooling function).

Looking in more detail at the function of respiration in the animal kingdom, in *On Respiration*, Aristotle observes that hot animals require full-blooded lungs, in order to get the rapid refrigeration necessary to conserve the vital fire of the heart. Some animals that do not produce much heat, and therefore require little respiration, have bloodless, spongy lungs. Indeed, some cold animals, like frogs and turtles, need so little refrigeration that they can remain underwater for a long time, not breathing. But they will suffocate if held under too long.

Lungless animals, such as fishes, do not breathe, for them, water passing over the gills is sufficient to provide the necessary refrigeration. Hence, because nature does nothing in vain (in today's terminology, all structures have a function), no animal possesses both lungs and gills. He ridicules Diogenes of Apollonia, who held that all animals, even aquatic ones, live by means of breathing air, and die if it is removed. According to Diogenes, when fish discharge water through their gills, they suck air out of the water by means of a vacuum formed in the mouth. This is untenable, Aristotle reasons, because respiration is necessarily of two parts: Inhalation and exhalation of breath. How then do fishes expire? It is impossible that they exhale and inhale the air at the same time. Also, if fish can take air from water, then why not men, and, if fishes respire air, why do they die in air? He rather grandly concludes that previous writers were in error, because they had no acquaintance with the internal organs of animals.

Aristotle recognized the cetaceans, dolphins, and whales as a special case: Although aquatic and fish-like in outward appearance, yet they have lungs. They must therefore use the lung to respire air for refrigeration. "Hence they sleep with their heads out of water, and dolphins, at any rate, snore." If they are entangled in nets, they soon die of suffocation, owing to the lack of respiration. And he

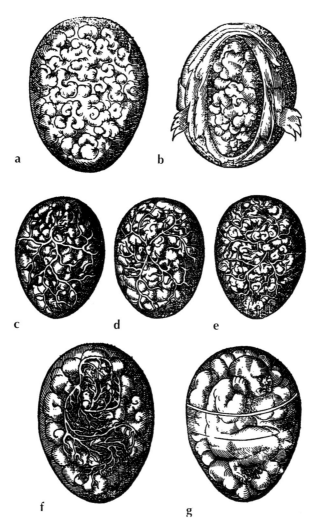

Fig. 4.4. Aristotle's conception of embryogenisis. (The illustration is from Jacob Rueff's *De Conceptu et Generatione Hominus* [1554] arranged by Singer [1957].) At (**a**) the menstrual blood is gathered in the womb; (**b**) shows the hot semen, with flames coagulating the blood; (**c**) note the first appearance of the blood vessels; (**d**) the first moving principle, the heart; (**f**) the "sketch" outline of the child; and (**g**) the child sculpted in more detail. (Reproduced with permission from Dover Publications.)

further notes that, when they feed in seawater, they swallow water with the food, and this is discharged through the blowhole.

The Senses

All sensations start and end in the heart, making the heart the common sensorium. In *On the Soul (De Amina)*, Aristotle declares that the senses work by receiving the sensible forms of things, without the matter. This operates rather like how a piece of wax takes on the impress of a signet ring without diminishing the iron or gold in the ring. As before, a good analogy can provide a sufficient explanation.

In *On the Senses and Sensibilities (De Sensu)*, he argued that the assertion of Plato and Empedocles—that vision is the result of light issuing from the eye as from a lantern—is absurd because, in the absence of light nothing can be seen. The objection that luminescent things, such as fungi and the eyes of dead fish, can be seen in the dark, is not valid, because these are exceptions, and they do not have "proper" color. To say that we cannot see in the dark because darkness quenches the visual ray, is idle: Although water might put out a fire, rain does not quench the light of the sun.

Movement

For Aristotle, the animal soul, the pneuma, is the material mover. Pneuma is localized in the heart, and is transmitted through the "sinews" (solid, dry, hard, and elastic tissues) to the limbs. Animal movement is produced by a mechanism similar to that of the string puppet, by the pulling and release of the sinews.

Reproduction

Women, said Aristotle, had less heat than men, therefore, they were less vital or lively (is he being humorous, a little Aristotelian joke?). The menstrual flow supplies the material for the new being. But the actual work is carried out by the hot semen. This semen is made in the testes during the last pepsis. On contact with the menstrual fluid, semen sets it like rennet "sets" milk, then gradually shapes the fluid into the infant, like a sculptor (Hall, 1969a). Another analogy was the painter who first sketches the outline, then applies more detail with color for the final product.

Aristotle's embryogenesis is illustrated in Fig. 4.4, where can be seen: (A) the menstrual blood gathered in the womb; (B) the hot semen, with flames, coagulating the blood; (C) the appearance of the blood vessels; (D) the first moving principle, the heart; (F) the "sketch" outline of the child and (G) the child sculpted in more detail. This process of embryological development, from the simple to the complex, was called epigenisis. As we shall see, the doctrine fell out of favor in the seventeenth century, but was revived in the eighteenth.

Kidney and Excretion

Aristotle discussed the function of the kidney, in *On the Parts of Animals*. He believed, probably from personal observation, that kidneys are not found in fish or birds, and therefore deduced that the kidneys are not essential for life. His error is understandable. All vertebrates, of course, have kidneys, but in some, such as birds and fishes, the kidney is not a discrete, rounded mass, but is a diffuse organ. Further, he considers that, although not essential, the kidney's must have a purpose, and concludes, "The kidneys are not present for necessity in animals but have the functions of perfecting the animal itself." Their specific function was to purify the blood. "Kidneys are present to serve a good purpose: that is to say, their particular nature enables them to cope with the residuum (urine) which collects in the bladder *On the Parts of Animals*."

More or less correctly, the kidney was thought to separate the liquid wastes from the blood, and to send them to the bladder, but, how this separation occurred was not clear. A later, widely held filter model, in which wastes passed from the blood to the urine across a dividing membrane in the kidney, was attributed to Aristotle, but without good evidence (Marandola et al., 1994). The kidney had another, but lesser, function, a mechanical one: to provide an anchor for the lower blood vessels. To this end, the kidney, liver, and spleen had similar roles, to act like rivets in securing the blood vessels in their proper positions (otherwise, they would collapse in a heap in the lower body).

Aristotle's thoughts on the functioning of the heart, respiration, the brain, assimilation of food, excretion, and reproduction dominated Western science, such as there was, for the next 2000 years. But his lasting importance for biology and, indeed, all science is his method of finding knowledge. His approach is to seek a natural answer by observation, reasoning, argument, and detecting inconsistencies and contradictions in arguments. It is not a matter of opinion—you believe this and I believe that. On the contrary, it is a matter of demonstrating the error in wrong explanations.

However, the dependence on logical argument, the dialectic of thesis and antithesis, to resolution, was both a strength and a weakness. Scientific debates were conducted like arguments in a court of law, probably, in some cases, before an audience. The danger of the dialectic is that, when the answer to the supposed purpose is arrived at, inquiry stops.

5 The Alexandrian Period

The Age of the Experiment and the Textbook

Greek science is collected, codified, and systematized in the Alexandrian Museum, a state-subsidized collection of scholars. Initially, important anatomical and physiological discoveries were made, and physiology was established as a science. But, over time, there was a gradual shift from science toward mysticism. As the museum declined to extinction, most of what was known of Greek science was lost.

Early in his short career as world conqueror, Alexander the Great (356–323 BC) captured Egypt, and, about 330 BC, he founded the city of Alexandria. His favorite general, Ptolemy, took control of Egypt when Alexander died in 323 BC, and founded the Ptolemic dynasty which ended with the death of Cleopatra and her son, Ptolemy XV Caesar, in 30 BC. Ambitious to establish Alexandria as the capital of Hellenistic culture, Ptolemy I, called Ptolemy Soter (the savior), established two famous libraries, the Alexandrian Library, associated with the temple of Zeus, and the Museum Library, connected with the temple of the muses (hence "museum"). Demetrius of Phaleron, one of Aristotle's earliest students, and classmate of Alexander the Great, was commissioned to oversee the collection of Greek scholarship and the translations of foreign knowledge into Greek. The center of Greek science, in consequence, moved from Athens to

Alexandria. Although the museums were set up with high priests at their heads, their original mission was secular rather than religious. They were to assemble, copy, organize, and disseminate the knowledge of the known world: Greece, Egypt, Macedonia, Babylonia, and beyond. They were to transmit this knowledge through teaching, and they were to produce new knowledge through study and research (Farrington, 1949).

Later accounts describe the growth of the museum complex into a most imposing assemblage, with courts, botanical gardens, and a zoological park containing exotic animals from the furthest parts of the known world. At its center was a great hall and a circular, domed. dining hall, with an observatory in its upper terrace. These buildings were surrounded by classrooms and demonstration rooms, including dissection laboratories. The museum had a faculty of 30–100 permanent schol-

Fig. 5.1. Greek illustration of human operation (or perhaps dissection). (From the National Library of Medicine.)

ars, and accommodation for visiting scholars. Salaries were funded from state revenues, and by direct donations from Ptolemy I and his filial successor, Ptolemy II. At its peak, the Library was reputed to contain more than one half million volumes. Two thousand years ahead of its time, it functioned as the first state-funded university.

Early Alexandrian scientific accomplishments were astonishingly rich. In astronomy, the complicated Ptolemaic system, which accounted for the movement of the heavenly bodies around the earth, was the dominant model for more than 1000 years. The mathematics of Archimedes and Euclid were taught in high schools until quite recently, and Alexandrian advances in anatomy and physiology were not surpassed until the seventeenth century.

Undoubtedly, a major impetus behind the latter was the enthusiasm for human anatomy studies engendered by Ptolemy's granting permission to dissect human corpses (Fig. 5.1). In Greece, as elsewhere, cutting the human body into parts had been strongly forbidden, for religious reasons. Human dissection was seen as a horrific desecration: The mutilated body would not be able to enter into the afterworld. Consequently, all human anatomical knowledge had come from inference from animal dissection. But now, for the first time, it was possible for Greeks to delve into the unexplored territory of human anatomy, and this new

knowledge stimulated new investigations into function (French, 1978b).

However, some thought that the early human anatomists went too far, and there were charges that, in their eagerness to discover the secrets of the human body, some were resorting to vivisection. Writing 250 years later, the Roman encyclopedist A. Cornelius Celsus (ca. 25 BC–50 AD), stated bluntly that Alexandrian anatomists performed human vivisection. In his *De re Medicina* (the eight books of which were the only ones to survive his huge encyclopedia on many subjects), he writes that, in order to know how the body actually is:

" … it is necessary to dissect the bodies of the dead and examine their viscera and intestines. Herophilus and Erasistratus, did this in the best way by far when they cut open men who were alive, criminals out of prison, received from kings. And while breath still remained in these criminals, they inspected those parts which nature previously had concealed, their position, color, shape size, arrangements, hardness, softness, smoothness, relation, processes and depressions of each, and whether any part is inserted into or is received into another."

But he excuses them, adding, "nor is it cruel, as most people maintain, that remedies for innocent people of all times should be sought in the sacrifice of people guilty of crimes, and only a few at that"

(Longrigg, 1988). Given the times and circumstances, such accusations may have been true, and, indeed, the practice of experimenting on prisoners receives support from some conservative quarters in the United States even today. But, in any event, about 40 years later, the religious prohibition was reasserted and human dissection was once more forbidden, not to be practiced again until the Renaissance.

The best of Alexandrian biology is exemplified by the outstanding achievements of the two above-mentioned scientists, Herophilus and Erasistratus.

Herophilus of Chalcedon (325–255 BC), the most famous of the Alexandrian anatomists, certainly performed human dissection. He seems to have explored this rich new territory of the human body with great enthusiasm, describing, in at least 11 treatises, human anatomy, in much greater detail than ever before. Unfortunately, none of his books have survived, but a long list of important anatomical findings is attributed to this investigator (Longrigg, 1988). He gave an accurate account of the digestive system, and discovered and named the duodenum. He was the first to dissect out the four tunics (or coats) of the eye, which he named the cornea, retina, choriod, and iris, and he also described the vitreous humor (Wiltse and Pait, 1999). Originally, and without any apparent precedent, the structure of the ovaries were compared to the testicles, and the epididymis was located and named. Herophilus provided accurate descriptions of the livers of different animals, including humans, and commented on species differences in size and shape. In his account of the vascular system, he distinguished, for the first time, between the structure of the arteries and veins, noting that the arteries had tunics six times as thick as veins.

But Herophilus's most impressive and original investigations are on the nervous system. From anatomical studies of the human brain, he distinguished between the cerebellum and the cerebrum, and specified the cavity of the large cerebellum (the fourth ventricle) as the seat of intellect (Sullivan, 1996b). From dissection, he was able to differentiate between tendons and nerves, which,

up until that time, were thought to have the same "cord" nature, and he was the first to further differentiate between cranial and spinal nerves. Even more, six pairs of cranial nerves were identified and named: the optic, occulomotor, trigeminal, facial, auditory, and hypoglossal (Wiltse and Pait, 1999). He traced the optic neuron from the brain to the eye. And, finally, through careful dissection, he revealed that the origin of all nerves lay in the brain and spinal cord, and that they terminated in other organs. Perhaps through experiment, he distinguished between sensory and motor neurons, and proposed that the motor nerves served as conduits for motion-giving pneuma that arose in the brain.

Herophilus appears to have been a person of independent mind. Disagreeing with his old teacher Praxagoras, he held that arteries contained blood, not pneuma, and, contradicting Aristotle, he argued that the brain was the seat of intelligence, rather than the heart. He maintained that the pulse in the arteries derived from the heart, not from the ebb and flow of pneuma. Using a portable water clock as a timekeeper, he was able to compare pulse rates of healthy and ill people. This new technology encouraged him to construct a complex classification based on pulse size, strength, rate, and duration, to be used as a diagnostic tool. This useless, but patient-impressive, system, of pulse examination continued to be practiced until recent times.

The findings of Erasistratus of Chios (310–250 BC), the most famous of the Alexandrian physiologists, are also known only from secondary sources. A prolific author, he is recorded as having written works on anatomy, pathology, hemoptysis, fevers, gout, dropsy, and hygiene. His fame, however, rests on his complex and comprehensive system of physiology. He was interested in how the body worked, and, to this end, he was involved in human dissection and animal experimentation (and perhaps also human vivisection).

Erasistratus believed that every organ was supplied with a threefold system of "vessels" the veins, arteries, and nerves, which divided and subdivided beyond vision, vessels so fine that they could only be comprehended by reason (Singer, 1957). He located the origin of both the veins and arteries in the heart, and of the nerves coming from the brain. This vessel system was at the base of his physiological theories.

Erasistratus rejected the humoral theory, replacing it with an elaboration of pneuma which devolves into animal and vital spirits. The original pneuma came from the inspired air, and passed to the left side of the heart, where it is changed into vital spirit, and it is then transmitted via the hollow arteries to the rest of the body. This vital spirit was the source of the innate heat of the body, and supported digestion and nutrition. In the ventricles of the brain, vital spirit was converted into animal spirit, which was conveyed, by the hollow nerves, to different parts of the body. Animal spirits endowed the individual with life and perception and motion.

Like Praxagoras, then, Erasistratus maintained that the arteries contained only pneuma, not blood. He has a saving explanation to the objection that blood gushes out of a wounded artery. When the artery is damaged, its pneuma escapes, causing a vacuum, and, as nature abhors a vacuum (from Strabo), blood is drawn into the arteries from the veins, through fine branch connections that are ordinarily closed. He apparently had guessed the existence of the capillary system (Singer, 1957). Erasistratus gave the first accurate description of the heart. He discovered and named the bicuspid and tricuspid valves, and suggested that the heart acted as a pump (Longrigg, 1988). He appears to have conceived that the heart operates as a two-stroke pump, combining suction and pushing, to simultaneously move the two different kinds of fluid, blood and pneuma. But the flow is one way, with pneuma and blood moving slowly to the peripheries, and nutrient blood being consumed by the tissues.

Erasistratus also had a mechanical explanation for digestion, absorption, and nutrition. After being broken up in the mouth, food is passed to the stomach, where it is mashed to a pulp, or "chyle," by peristaltic action of the gastric muscles (he described the action of the epiglottis in pre-

venting food and drink going into the lung). The chyle is squeezed through the walls of the stomach and the intestine into the blood vessels communicating with the liver, where it is transformed into blood, which then passes to the right ventricle of the heart. The blood is then pumped into the lungs through the vein-artery (pulmonary artery), from where it is distributed as nourishment for the rest of the body through the all-reaching venous system. Some tissues, however, such as the brain, liver, and lung, in addition, also had a deposit of nutriment, called "parenchyma," a term still used today. As we shall see, this theory of nutrition, and its variants, appear and reappear in later history.

Blood as food formed the basis of Erasistratus's theory of illness. An overabundance, or flooding, of the veins with blood (*plethora*), was thought to be the main cause of disease. As the plethora increased, the limbs would swell, there would be local inflammation and fever, and organ function would be mechanically impeded. In consequence, the rational treatment is to stop nutrition through starvation, or to bleed off excess blood (phlebotomy). Phlebotomy remained a major medical treatment until the nineteenth century.

Erasistratus appears to have carried out one of the first known experiments in quantitative physiology. It is recorded that, in order to prove that living creatures give off invisible discharges, he weighed a bird, then held it in a vessel without food for some time, then weighed it, together with all the excrement that had passed. This experiment was repeated in a greatly elaborated scale in the eighteenth century by Sanctorius (*see* Chapter 12).

Like Herophilus, Erasistratus had a special interest in the structure of the brain, and described many detailed features. For example, he succeeded in tracing nerves to the interior of the brain. Erasistratus agreed that the fourth ventricle in the cerebellum, was the seat of intellectual activity. This double authority confirmed a ventricular theory of brain function that was to dominate scientific thought until the eighteenth century. He likened the convolutions of the brain to the coils of the small intestine (a concept of great tenacity) *(see Fig. 10.13)*, and compared the extent and number of cerebellum convolutions in different animals, including hares, stags, and humans. Imaginatively, he suggested that the higher intelligence of humans might be attributed to the greater number of convolutions in the human brain.

Of course, other Alexandrians made original discoveries in biology. The mathematician, Ptolemy, proposed a theory of vision that was to prove most influential. He accepted the view that rays shining out from the eyes transformed the intervening air into an instrument of sensation, which allowed direct apprehension of the object, which the nerve transmitted to the brain. But he provided a geometric analysis of vision in terms of visual pyramids formed by the rays connecting with points on seen objects, which gave, for the first time, an explanation of three-dimensional perception (Gross, 1999).

However, Alexandrian science was not sustained, and, after a brilliant start, rapidly declined. Human dissection stopped, following the loss of protection and provision by the first Ptolomies, and interest in scientific discovery and rational explanations waned. In its place, the focus of attention shifted to looking for supernatural causes of events, and for signs of divine revelation. The library became concerned with building up a collection of what has been called hermetic literature, mystical works attributed to the god Hermes.

Without doubt, part of the reason for this flight from science was the instability of society and the institute. The library had a rough history over the next 600 years, being destroyed and rebuilt, with an overall gradual decline in intellectual activity. According to Livy, for example, over 40,000 volumes, housed in grain depots near the harbor, were incinerated when Julius Caesar torched the fleet of Cleopatra's brother. And there is the legend of Hypatia, a fifth-century scholar and mathematician of Alexandria, being dragged from her chariot by an angry pagan-hating mob of monks, who flayed her alive, then burned her upon the remnants of the old Library (Osler, 1921). Nothing remains of the fabric of the library building, but, by great luck, vital classical texts were translated into Arabic and Hebrew, in which they were preserved long after copies were lost during the Middle Ages in Europe. These translations account for almost all of our knowledge of Greek science.

6 The Roman Period

CONTENTS

Compared to the Greeks, Roman science was applied and practical, with little interest in theory and speculation about the true nature of things. Great uncritical treatises were written by writers such as A. Cornelius Celsus and Pliny, consolidating current biological knowledge (and error). Lucretius speculated on the nature of life but substantial advances were made by Galen, particularly in anatomy and physiology. Galen's physiological system of spirits was treated almost as dogma until its overthrow in the Renaissance.

As Rome assimilated Greek culture, the Greek tradition in science was transferred to Rome. But the Romans were not interested in speculations about how the world worked; they wanted practical information to protect the individual and advance the state, such as was provided by engineering and as promised by divination. With the exception of Galen, the Romans provided little new in physiology, and only a few Romans influenced later biological thought (Farrington, 1944: Hall, 1969a).

CELSUS

Aurelius Cornelius Celsus (ca. 25 BC–50 AD), was an educated nobleman with an amateur interest in medicine. During the reign of the Roman Emporer Tibereius (14–37 AD), Celcus compiled a general encyclopedia of agriculture, medicine, miliary strategy, philosophy, law, and rhetoric, of which only the eight books on medicine, *De re medicina* eventually survived. During the reign of the Roman Emporer Tiberius (14–37 AD), Celsus compiled a general encyclopedia of agriculture, medicine, military strategy, philosophy, law, and rhetoric, of which only the eight books on medicine, *De re medicina* eventually survived. This voluminous treatise on medicine and surgery, is essentially a compilation of contemporary medical knowledge, with little in the way of original thought (Sullivan, 1996c). In some ways, it was a practical handbook, with chapters on such topics as diseases to be treated by diet and lifestyle, and diseases that could be treated by surgery and medicine. The treatise was lost in medi-

Fig. 6.1. Galen wished to demonstrate universal harmony in body structure by proving that "the organs are so well constructed and in such perfect relation to their functions that it is impossible to imagine anything better." Galen conducts physiological experiments on a live pig tied to a table. (From Galen Venetiis, 1586 [National Library of Medicine].)

eval times, but was later rediscovered by Pope Nicholas V (1397–1455) (Castiglioni, 1958). *De re medicina* was so esteemed that it was printed in Florence in 1478, making Celsus the first classical medical writer to appear in print. Eight books in all were published, including a history of medicine that discussed human vivisection in Alexandria, which we mentioned in Chapter 5. Indeed, much of our knowledge of Alexandrian medical science comes from Celsus. His clear and straightforward Latin made the books popular during the Rennaisance and he remained an enduring influence in surgery until more recent times.

PLINY

Caius Plinius Secundus, known as Pliny the Elder (23–79 AD), produced or dictated an enormous 37-volume *Historia Naturalis*, an encyclopedia of science of the period, drawn, as he claimed in the preface, from about 2000 works, almost all lost, by 146 Roman and 326 Greek authors. The book was divided into eight sections arranged to cover all knowledge (Singer, 1958): Book 1, Introduction; Book 2, Cosmology; Books 3–6, Geography; Book 7, Anthropology; Books 8–11, Zoology; Books 12–19, Botany; Books 20–32, Medicine; and Books 33–37, Mineralogy and Art. This massive undertaking was described by Edward Gibbon

in his *Decline and Fall of the Roman Empire* (published 1776, 1781, and 1788 in 6 volumes) as "that immense register where Pliny has deposited the discoveries, the arts and the errors of mankind." The motto of the *Historia* is "Nature serves man," and natural objects are described in terms of their usefulness to humans. As we shall see, this concept was still the mode 1000 years later, in the medieval bestiary. Indeed, Pliny's *Historia* was the most authoritative book on natural history until the end of the middle ages. But, as Gibbon indicates, it is an uncritical collection of facts and fables. For example, Pliny describes, without any qualifying distinctions, the natural histories of the lion, unicorn, and the phoenix, and exotic human races such as the Psilli, whose bodies secreted a poison deadly to snakes, or the Sciopodes, who have a single enormous foot. He provides curious remedies for illnesses: "On pulling off the head of a blatta (cockroach) it gives forth a greasy substance, which beaten up with oil of roses, is said to be wonderfully good for affections of the ear," and "the nature of jaundice is yellow and if a jaundiced person looks on a golden oriole he recovers but the bird dies."

Pliny, however, paid the price for his insatiable, but indiscriminate, curiosity. He was in command of the Roman fleet at the time of the eruption of Vesuvius that destroyed Pompeii, and was so eager

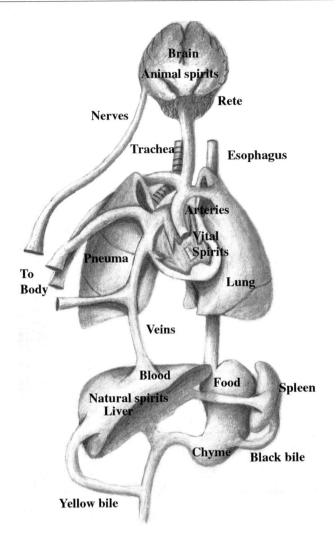

Fig. 6.2. Model of Galen's triadic plan of physiology. A detailed commentary is in the text. (Drawn by Thompson; from Peter Lutz, 2001.)

to see what was happening that he rashly landed, and disappeared in the cloud of ashes.

A minor figure, Rufus of Ephesus (98–117 AD), wrote books on anatomy and diseases. He appears to have been the first to describe the human liver as having five lobes. This is true of the pig, but not the human, but the error was repeated by Galen, and was believed to be true, until the time of Vesalius.

GALEN

For the later development of biology, Galen (ca. 129–210 AD) is by far the most important of the Romans. Galen, a Greek, was born at Pergamum,

and studied Greek medicine at Alexandria. He received extensive practical experience as a surgeon to gladiators, and became medical director in the training school for gladiators at Pergamum. His reputation was so high that he was appointed physician to the emperors in Rome. Galen was deeply interested in the structure and workings of the human body, and was a prolific writer. About 100 of his treatises are still extant, having been preserved in Arabic translations (*see* Chapter 8). Many had practical titles, such as *On the Usefulness of Parts of the Body*, *On the Bones*, and *Anatomical Exercises*.

In philosophy, Galen was influenced by the Stoics, who held that Fate determined all, all things were bounded by rule, and were ultimately determined by forces acting wholly outside the body. The universe interacted with humans and the study of this interaction—the influence of heavenly bodies on human fate and health or astrology—was accepted as a serious discipline until comparatively recent times. His views were fundamentally teleological, and in the book, *Uses of the Parts of the Body*, he approvingly expounds Aristotle's dictum that nature makes nothing in vain. However, his studies had a higher purpose than merely serving human health. Galen vas very influenced by the concept of the harmony of nature, and looked for harmony in the microcosm of the body. His mission was to demonstrate universal harmony in body structure, by proving that "the organs are so well constructed and in such perfect relation to their functions that it is impossible to imagine anything better" (Osler, 1921).

Galen's physiology was grounded on his anatomical studies, which were considerable, but, since human dissection was prohibited, he extrapolated from animal to human anatomy. His favorite animal subject was the pig (Fig. 6.1), but dogs and cattle were also used. As a result, some infamous errors in human anatomy were accepted as fact (or even dogma), right up until the time of Vesalius. For example, the human uterus was given two long horns, as in the dog, and the liver was divided into many lobes, as in the pig. Considering that human skeletons were available, his errors on human bone structure were even more extraordinary. For example, his human pelvis was flared like that of the ox.

Galen organized physiology on what was called the triadic plan: the three organs, (liver, heart, and brain); the three vessels (veins, arteries, and nerves), and the three spirits (natural spirit, vital spirit, and animal spirit). Each organ is compared to that central part of a tree which obtains materials from the roots (vessels), and on changes and sends them to the branches (vessels), to other parts of the body. The veins sprout

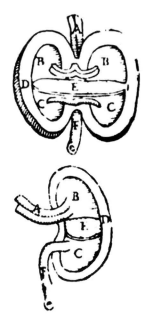

Fig. 6.3. The filtering kidney, according to Galen. The kidney was thought to consist of two chambers, an upper and lower, separated by a membrane, the colatorium, containing minute pores. Blood was supplied to the upper cavity from the renal vessels; it was purified as lighter and noxious fluid was filtered off by the colatorium. The filtrate was collected in the lower cavity, and conducted to the bladder via the ureters. (From Vesalius's *Fabrica*, 1543.)

from the liver, the arteries from the heart, and the nerves from the brain. Each organ was associated with one of the three adaptations of aerial pneuma, the breath of the cosmos. The first adaptation occurred in the liver, where the pneuma became natural or nutritive spirit. In the second adaptation, vital spirit was produced in the heart, and, in the third adaptation, which took place in the brain, the pneuma became animal or soul spirit (Fig. 6.2). The elements were used to build an imaginative and comprehensive system of nutrition. A basic premise was that every part of the body has its proper food, which it attracts and assimilates to its own substance. The liver, heart, and brain had special roles to play in the economy of the body (Siegel, 1969).

The Liver

Food is broken or concocted into chyle in the stomach, then moves through the intestine through the

Tabule

Fig. 6.4. Chart for the examination of urine (uroscopia). A wheel of urine flasks is depicted, each with an accompanying analysis of their contents. The physician sits in the center making his diagnosis. (From Ulrich Pinder, 1506 [National Library of Medicine].)

pyrolis. The liver attracts this nutritive chyle, which is drawn up through the portal system. In the liver, the chyle is elaborated into venous blood and imbued with natural spirit, thereby changing it into the liver's own nature, congealed blood (Wear, 1977). The resulting dark heavy blood, which flows slowly in the veins, serves for the general nutrition of the rest of the body. Drawn by the "attractive force" of the hungry tissues, the blood replaces or restores dead or dying tissues. Nutritive blood is distributed, in a closed venous system, by the branches of the venæ cavæ, one supplying the head and arms, the other the lower extremities; a small amount of blood is

delivered to the right chamber of the heart. One of the waste products, black bile or melancholy, is attracted to the spleen, from where it passes along a special (illusory) duct to the stomach (Fig. 6.2). But even black bile had a useful role: It compelled the stomach to draw up and clasp the food tightly, until it is completely concocted (Wear, 1977). To the objection that no connection between the spleen and the stomach had been seen, Galen answered there must be such a vessel: It was necessary for the perfect system he was describing. The spleen–stomach duct was accepted as an anatomical fact, until Vesalius.

The Heart and Blood Vessels

The heart was palpably the hottest part of the body, and was the source of the body's heat, produced by a sort of combustion, a process similar to a lamp burning, in which the blood is like the oil of a lamp, the heart a wick, and the "sooty vapors" engendered by this burning, pass up the pulmonary artery, and are released by the lungs (Fig. 6.2). Some venous blood, "the thinnest portion," is drawn into the left ventricle of the heart, the transfer taking place through tiny holes in the septum that separated the right chamber from the left (Galen imagined a three-chambered human heart). Here, the blood receives vital spirit, which has come from the lungs via the pulmonary artery, and is transformed into a thinner, brighter, warmer blood, the arterial blood. Air from breathing also fans and cools the cardiac flames. The warm arterial blood ebbs and flows in the arteries, which distribute vital heat to all parts of the body. Galen noted that it was not possible to observe the terminations of the interventricular pits, "owing to the smallness of these and to the fact that when the animal is dead all the parts are chilled and shrunken" (Osler, 1921). The cardiac septal pits were one of Galen's most stubborn errors, and were firmly believed in up until the seventeenth century. Galen believed that the dilation was the active phase in the heart's motion, not contraction. On dilation, blood is drawn in, like the action of a bellows, and moved out on contraction. Breathing and heart beat were different aspects of the same phenomenon, and the pulse was the arteries dilating to receive arterial blood.

Galen described an experiment to show that arteries contain blood, and not air. His procedure was to expose a length of artery in a living animal and tie off a portion with two ligatures. On cutting the isolated portion, he said that it will be seen that blood, not air, flows out. He deduced (correctly) that there must be fine (to him, invisible) connections between the arteries and veins: "If you kill an animal by cutting a number of its large arteries, the veins as well as the arteries become empty of blood.

This could never occur if there were not anastamoses between them" (Galen, 1916). In this exsanguination, venous blood must have flowed into the arteries.

The Kidney

In order to demonstrate that the urine was made in the kidneys, and not in the bladder, he observed that, if the ureters are ligated in a live pig, they will fill to the point of bursting. Removing the ligatures causes the bladder to fill with urine. In order to demonstrate conclusively that urine could not flow from the bladder to the kidney, he recommended ligating the penis and squeezing the full bladder: No fluid goes back into the ureters.

He speculated on the function of the kidney as a filter, an idea that was further developed by his followers. In this scheme, the kidney was thought to consist of two chambers, upper and lower, separated by a membrane, the colatorium, containing minute pores (Fig. 6.3). Blood was supplied to the upper cavity from the renal vessels; there it was purified, as lighter and noxious fluid was filtered off by the colatorium. The filtrate was collected in the lower cavity, and conducted to the bladder via the ureters. This was the accepted explanation of kidney function for the next 1000 years. It inspired an elaborate ritual of "uroscopia," whereby the physician determined health, illness, and treatment based on the color, smell, and taste of urine (Fig. 6.4).

In the sixteenth century, the great Vesalius scoffed at the filtration kidney as having no anatomical basis. Ironically, in the nineteenth century, it was finally demonstrated that filtration is indeed the primary method of urine production in the kidney.

The Brain and Sensation

The light, brighter, arterial blood receives animal spirits in the brain. This process takes place in a collection of fine blood vessels at the base of the brain, the rete mirabile (another long-enduring anatomical error; unlike ruminants, human brains do not have a rete mirabile). This fluid, which provides for movement and sensation, is stored in the brain ventricles, from where it is distributed to the rest of the body through the nerves, which "spring from the brain," and which have inner parts like the pith of a tree.

Galen performed a number of ingenious experiments to investigate organ function. For example, it was believed that that speech came from the chest, and that its source must be in the heart. Since we cannot speak without thinking, the heart must be the seat of mental activity. This he refuted by experimental demonstration, by cutting the recurrent laryngeal nerve, which can be traced to the base of the brain, which stopped dogs barking and pigs squealing. The origin of vocalization, he concluded, lay in the brain, not in the heart.

Galen was also something of a moralizing philosopher. He wrote in a short treatise, that "the Best Doctor is also a Philosopher . . . It is not possible to be successful in science and medicine unless one is hard-working; and it is not possible to be hard-working if one is a drunk, a glutton, or excessively addicted to sex: in short a slave to belly and genitals" (Sullivan, 1996c). Criteria, alas, that would exclude many (in)famous scientists of later times.

The Galenian system of physiology was an extraordinary intellectual feat. It was an elaborate, complex, difficult-to-understand, but highly intellectual schemata accounting for the working of the body. Most important it strived to link structure with function. It made sense of the anatomical structures then known to exist, and used them in a rational account of respiration, nourishment, and body heat. It provided satisfactory answers for a long time to come, and was the core of accepted medical practice. For example, good health required a proper balance in the three spirits, a doctrine still reflected today, when we speak of being in good spirits or of being high-spirited.

Galen believed that he was revealing a perfect design, and clearly felt free to improvise structures where he thought it necessary to complete his system of physiology, such as the ventricle pores. He compared the vessels to irrigating canals and gardens, and wondered at the goodness of the design that ensured organs "neither lack a sufficient quantity of blood for absorption nor be overloaded at any time with excessive supply." This is the sort of argument that present-day creationists continue to make. Indeed, Galen believed that discovering this harmony was a sublime task, since it was revealing evidence for the existence of God, the Great Designer. According to Singer (1958), this approach gave Galen a special appeal to the Christian point of view, and was the reason why more of Galen's work was preserved than any other pagan writer. Indeed, Galen was called the "Prince of Physicians" in the Middle Ages, and to contradict the authority of Galen was almost heresy (Castiligioni, 1958).

An interesting insight into the biological knowledge of the educated Roman of the fourth century is provided by a letter written by St. Basil (330–379) on the nature of the ant. He scoffs, how could the heretic, Eunomius, comprehend anything of the nature of God when he knows nothing of the nature of a trivial ant, and asks: "Does the ant live by drawing breath? Are her joints pulled and drawn by sinews and tendons? Is the setting of the sinews controlled by a system of muscles and glands? Does the marrow in her dorsal vertebrae extend from her occiput to her tail? Doth this spinal marrow, by means of a covering of sinewy membrane, give motive powers to her limbs? Hath she a liver, a gall bladder besides her liver, kidneys, a heart, arteries and veins, membranes and cartilages?" (Brooke, 1930).

But, compared to the Greeks, Roman science was applied and practical; there was little interest in theory and speculation about the "true nature of things." The Romans mostly confined themselves to the questions: Is it useful? Does it work? There was correspondingly little advance in new knowledge. But the Romans excelled in application, in practical enterprises, especially in engineering and architecture, to protect and control the empire. Sanitation was developed to an unheard-of degree, with massive aqueducts to bring fresh water, sumptuous public baths for recreation and to cleanse the body, and an elaborate system of sewers to wash away waste. Indeed, the collapse of the Roman Empire was followed by a rise of filth in the West.

7   The Middle Ages

Following the collapse of the Roman Empire, interest in science vanished in the West. Some texts, however, were preserved in monasteries, especially in Ireland and Northern Britain. As illustrated in the great and popular books called bestiaries, the perceived purpose of nature was to serve humans by providing food, labor, or moral lessons.

After the fall of the Roman Empire in the fifth century, almost all that was known in science was lost, and for the next 1000 years, no important new discoveries were made. However, we know that science re-emerged and flourished in the sixteenth century, so clearly the intellectual ground must have been in some way prepared during the Middle Ages for the sixteenth century renaissance in science. It follows that what happened in the intervening Middle Ages was actually quite critical to this rebirth of science.

In 313 AD emperor Constantine the Great (306–337 AD) issued the Edict of Milan, granting Christianity full legal equality with all other religions in the Empire. Only 70 years later, in 392 AD, Emperor Theodosius made Christianity the sole state religion of the Roman Empire, ordered all pagan temples closed, and prohibited pagan worship, on pain of treasonous death. As Gibbon so eloquently argues in his *Decline and Fall*, Roman regimes were indifferent to the religions and beliefs of their subjects, as long as they did not oppose the official Roman religion. Christianity had, however, revealed the one true God, and proceeded to identify all other beliefs as error at best, or works of the devil at worst, to label their adherents as pagans or idolaters, and to stamp them out vigorously. There was one Truth, but even that was argued about, with no tolerance shown for other opinions (called "heretical"). The early fathers of the church, such as Origen of Alexandria (ca. 185–254) and St. Augustine (354–430), formulated an elaborate but self-consistent intellectual construct based on the religion of the Bible and held together by Greek logic, which described in detail the spiritual relationships between God and humans, and humans and nature.

Origen argued that Greek philosophy was not inherently good or bad, and that it depended on how it was used by Christians. The Greek philosophers had produced a system of natural reason that could serve to reveal truth. Science and philosophy could be used as "handmaidens to theology," to help the understanding of Holy Scripture. However, the early Christian philosophical outlook was characterized by an indifference or hostility to speculative philosophy or science as missing the whole point of existence, which is the affirmation of the one true God and the discovery of his divine rules for the devout life, in order to achieve salvation and eternal life. All else is vanity.

Meanwhile, Christian Rome was overwhelmed by wave after wave of invading pillaging, pagan hordes (Goths, Vandals, and Huns), and, by the fifth century, the Roman infrastructure in Western Europe collapsed completely, and classical knowledge was almost extinguished. The next several hundred years, the Middle Ages, are also known as the "Dark Ages." There are few written records of this period, since writing was despised as an activity not fit for warrior kings and barons. Fortunately, the skill of writing was maintained in isolated Christian monasteries that were scattered throughout the pagan territories.

The monastic movement was a consequence of the perceived worthlessness of earthly pleasures and the external political and economic chaos. The monks renounced the vanities of the world, to live lives devoted to prayer and contemplation, removed from fleshly temptations. The earliest practitioners were individual hermits who lived lives of holy isolation, such as the anchorite St. Simon Stylites, renowned for spending years atop a pillar in the desert, refusing to wash or even to remove maggots from his rotting flesh, reputedly saying to the maggots, "eat what God has given you" (Haggard, 1928). Communities of monks were established, such as the many small, isolated Celtic monasteries of Ireland and northern Britain (Cahill, 1995), and some gradually transformed into larger organizations, such as the massive Benedictine monastery at Monte Casino in middle Italy. The life of a monk was governed down to the minute by strict rules of conduct.

As the monasteries thrived, the monks became more confident and evangelical, and set out as missionaries to gradually reconquer pagan Europe for Christianity, and to spread the written word.

One of the monastic duties was to preserve and transmit sacred writings and commentaries, but a few classical manuscripts were also copied. For example, Bede of Jarrow (673–735) produced manuscripts commenting on the natural world, which were chiefly founded on Pliny's *Natural History*. Of the writings of Aristotle, the logical works that were translated by Boethius (480–524) survived. Indeed, it has been claimed that, without the scholarly work of the Irish monks, most of the classic Latin writings would have been lost (Cahill, 1995). However, these parchment books required much time to write and illustrate, and were scarce and very valuable. Some increase in productivity was achieved through a division of labor, in which the text was read to several scribes by a single lector, and a limner later added the illustrations (Fig. 7.1).

The books functioned more to conserve rather than transmit ideas. Since religion ruled all monastic thoughts and deeds, the monks were taught, and believed, that the universe was governed by a controlling mind, and that everything was capable of explanation, and that everything meant something. As far as the abundance and variety of living organisms was concerned, it was very simple: Nature was made by God for man, and all creatures fulfill a special purpose to that end. The purpose of animals is to provide food, clothing, and labor, to be living illustrations of sacred events, or to give moral lessons. For example, in his book *The City of God*, St. Augustine uses the examples of the continually burning, yet unconsumed, salamander and the smoldering Mt. Etna, to illustrate that God could make human bodies burn forever in Hell. In *On Christian Doctrines* he says, "The well-known fact that a serpent exposes its whole body in order to protect its head from those attacking it illustrates the sense of the Lord's admonition that we be wise like serpents." From his knowledge of God's great plan, Augustine deduced that human beings could

Fig. 7.1. Medieval scribe at his desk. (From White, 1954.)

not live at the antipodes (the opposite side of earth), otherwise they would not be able to see Christ descending from Heaven at the second coming.

Wonderful (in the sense of causing wonder) illustrations of animal moral lessons were provided by a great and popular book, the bestiary called the Physiologus, which was a compilation of medieval natural history, whose purpose was to show divine revelation in the book of nature. Its roots go back to mythology. The *Physiologus* was widely translated (in prose and rhyme) into most Middle Eastern and European languages, including Arabic and Icelandic, and it was the basis for many later bestiaries. Next to the Bible, bestiaries were the leading books in twelfth and thirteenth century England. According to Singer (1958), the last handwritten translation of the *Physiologus* itself was into Icelandic in 1724.

A few examples, based on White's (1954) excellent and entertaining translation of a twelfth century Latin bestiary, will provide some idea why this concept was so popular.

" ... The lion, the king of beasts, illustrates nobility: it disdains to kill sleeping prey and will wait till it awakes. The lion also shows mercy; it will spare the prostrate."

The crocodile (Fig. 7.2A) breeds on the River Nile.

"It is amphibious, is 30 ft long, armed with horrible teeth and claws. The hardness of its skin is so great that no blow can hurt the

Fig. 7.2. Illustrations from a medieval bestiary (White, 1954). **(A)** The crocodile: a creature so hard that stones would only bounce off its skin. **(B)** The hyena: a loathsome beast, robbing a coffin. **(C)** The pelican: a noble bird, first justly killing its disrespectful young, but then mercifully reviving them with its own blood.

crocodile, not even hefty stones which are bounced on its back. It is still by the light of day and does its evil under darkness at night. Hypocritical, dissolute and avaricious people have the same nature as this brute."

The illustration is pure imagination. The limner had obviously never remotely seen, nor even met anyone, who had actually seen, such a creature.

The hyena (Fig. 7.2B) devours dead bodies; "by nature it is at one moment masculine, at another feminine, and hence it is a dirty brute. It walks around houses by night and, in order to prey upon men it copies the sound of human vomiting (!), calling out men at night (to their horrible death)."

The pelican (Fig. 7.2C) (which was the most popular of all the beasts) "is excessively devoted to its children, but when they begin to grow up they flap their parents in the face with their wings, and the parents strike back and kill them. Three days afterwards, the mother pierces her breast and lays herself across her young, pouring her blood across their dead bodies. This brings them to life again. In the same way, the crucified Christ gives blood from his side for our salvation and eternal life, (and so on)."

The Almighty also revealed healing herbs through the "doctrine of signatures"— God had given healing plants special signs by which they could be recognized. The red juice of beet root was good for blood disorders; liverwort, a liver-like leaf cures liver diseases; eyebright, marked with an eye-like spot, remedies eye diseases.

However, as time went on, the concept of nature as a demonstration of God's glorious design became more elaborate. A good impression of the medieval picture of the complex plan of the universe can be obtained from the writing of Hildegard, prioress of Bingen. Hildegard (1098–1179) was born in Bokelheim, in rural the Rhineland. Her life spanned what has been called the "twelfth-century renaissance", an era in Europe of expanding population and growing new cities, of the rein-

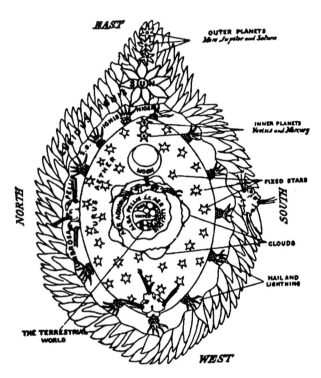

Fig. 7.3. Saint Hildegard's first vision of the universe (translated and modified by Singer, 1958). Reproduced with permission from Dover Publications.

vention of a money economy, of written law, and of government bureaucracy. Hildegard was a woman of extraordinary accomplishments. She was a prolific writer, and wrote poems, hymns, and works on natural history and medicine. She was a philosopher of considerable influence, and was consulted by, and advised, bishops, popes, and kings, including Henry II of England, Louis VII of France, the Holy Roman Emperor, and the Empress of Byzantium.

According to Singer (1917), Hildegard, as a typical medievalist, makes no distinctions between physical events and spiritual experiences. Her visions are equally valid to her sensory observations and serve moral truths. All are interwoven, a fusion of external and internal universe. Music was an important component of this universe for Hildegard, who, as a prolific composer, wrote more than 70 Gregorian chants and two musical dramas. After 1000 years of neglect, her music is enjoying a current revival.

Fig. 7.4. Hildegard's vision of the arrival of the unformed soul (Singer, 1958). The fetus appears to be fairly well developed before it receives the soul to become human. Reprinted with permission from Dover Publications.

Hildegard's book, *Scivias* (*Know the Ways of God*) ca. 1148, is her major work of visions. Each is described by a beautiful illustration with an

elaborate text. For example, she reveals her visions of Earth surrounded by spherical and oval zones that house the winds, clouds, planets, stars, and sun (Fig. 7.3). She pictures the unformed soul entering the unborn child (rather late in gestation) (Fig. 7.4), and the formed soul leaving the body after death (Fig. 7.5). Hildegard extends, in great detail, the concept of the relationship between the outer universe, the macrocosm, and the organization of the human body, microcosm, developing the basic premise that the body is ordered and unified, like the universe. The four elements (earth, water, air, and fire) were not seen as abstract principles, but as concrete elements of land, rain, wind, and sun, and health or goodness is a proper balance of nature without and nature within (Singer, 1958).

Physica, or *Subleties of the Diverse Quality of Created Things* (ca. 1158), deals with the natural world and its uses as cures for ailments. It consists of nine books: Plants, Elements, Trees, Stones, Fish, Birds, Animals, Reptiles, and Metals. Each item is categorized and its proportion of the four basic qualities—hot, cold, dry, and moist, and the medicinal value of the item or its parts—is described in detail, providing a curious mixture of fantastic and pragmatic natural history information.

A few examples will suffice, derived from Throop (1998). The whale has a fiery heat and watery air, and it feeds on fish. "If the fish in the sea were not eaten and so diminished, the multitude of fish would leave no passageway in the sea." "The ostrich is of such great heat that, if she were to keep her eggs warm herself, they would be burned up, and her young would not come forth. And so she conceals them in the sand." "When a bear conceives she is so impatient in childbirth that, in her impatience, she aborts before the cubs have come to maturity … seeing this, … the mother grieving, licks it, passing her tongue over its features until all its limbs are distinguished." The nicest description concerns the unicorn. "The unicorn is more cold than hot … it flees humans and so cannot be captured." But there is an exception: It can be captured by girls, but the girls must be nobles, not country

Fig. 7.5. Hildegard's vision of death and the departure of the formed soul. Angels to the left, demons to the right, heaven above (Singer, 1958). Reprinted with permission from Dover Publications.

girls, and must be in "the midst of adolescence. ... The unicorn loves them, because it knows they are gentle and sweet."

Her book, *Liber Divinorum Operum Simplicus Hominis*, completed around 1173, is a visionary text that deals with the origin of life, the physiology of the body, and salvation. The elements of the macrocosm are intimately related to the organization of the microcosm through elaborate parallels,

such as, for example, flesh equals earth or soil, vein are rivers, bones are rocks, and hair is grass. Parallelism is developed in great detail. "The firmament contains stars just as a man has veins that hold him together ... and just as blood moves in the veins and moves them with the pulse, so does the fire in stars move them, and emits sparks like the pulse."

In this book, Hilegard provides an overview of medieval medical science, including discussions on food and drink, rest and exercise, emotion and sex, and even orgasm. As detailed by Singer (1917), her elaborate justification of uroscopy calls upon parallelism. It is the practitioner's most reliable tool, because the urine reveals the state of the patient, just as a river reflects the weather.

> "As a river's waters change according to the winds and weather ... so also does the urine. While a person in a balanced, moderate state will have a still, balanced urine, a dangerous illness is like a great wind and shows itself in the urine like a tempest. Extreme fever is like a burning sun and in fever the urine is hot and burning. If the urine is white like poison or coagulated milk, and if, in the middle, it is like a cloud that is both dark and purple, then this is a sign of death, and the patient will surely die. For the urine is white like poison and similar to coagulated milk because the natural heat of the patient is about to leave. When the urine is dark, white, cloudy, and turbid in the middle it is because the melancholic humor has turned dark like a bruise; it is white because the patient is weak, and turbid because the patient has begun to excrete that natural but evil smoke which previously he had contained within (Singer, 1917)."

The increase in intellectual inquisitiveness at the end of the twelfth century is illustrated in Hildegard's writings, with the construction of more and more complicated schemes to account for the natural world. But they are based on the bits and pieces of classical knowledge that had survived, elaborated by imaginative fictions. In the absence of new information and fresh attitudes, scientific scholarship might easily have settled down to hundreds of years of sterile enquiry. However, the medical school at Salerno, Italy, was a harbinger of change to come. Legend has it that the school was founded in the ninth century by a Greek, a Latin, a Jew, and a Saracen (Castaligioni, 1958), symbolically representing the four sources of knowledge. It was more secular than religious, and reputedly had some female teachers, the most famous of whom, Trotula (though perhaps only a legend), was credited as the author of several manuals on obstetrics. In the tenth century, Hippocratic medicine (or what little was known of it) was taught to paying students, and by the beginning of the twelfth century, the school's reputation for scholarship blossomed, fed by the new knowledge coming from translations from Islamic science, the subject of the next chapter.

8   Islamic Science

While the candle of knowledge sputtered and was almost extinguished in the West, it shone with a new light in the East. The holy fire of Islam, which rose in the Middle East, rapidly spread throughout Asia, North Africa, and Europe. By the seventh century, the Arabs had conquered Persia, Syria, and Egypt, and the surviving Greek texts came into their possession. They captured North Africa, then Spain, which they held for almost 500 years until the expulsion of the Arabs by Queen Isabella in the fifteenth century.

From the beginning, Islam was a religion of the word, which valued literature. Inspired by the *Koran* to "Seek knowledge from the cradle to the grave, ... even as far as China" and the belief that "the ink of the scholar is more holy than the blood of the martyr," caliphs and notables subsidized translations of Greek, Hebrew, and Syriac science and philosophy texts. Translations of Aristotle and Galen into Arabic were critically influential for the later development of science. From the ninth to the eleventh centuries, the two capitals of the Arab world, Baghdad in the Eastern Caliphate, and Cordova, capital of Andalusia in the West, were among the most important centers of culture and civilization in the Western world. One of the finest translators, Honein (born 809, known in the West as Joannitius) translated into Arabic many of the works of Hippocrates, Aristotle, and Galen (Osler, 1921). The Arabic textbooks illustrated anatomical details such as the skeleton (Fig. 8.1), and the venous system (Fig. 8.2), and provided functional, rational, speculations on how the body works that were vastly more sophisticated than the naive commentaries in the contemporary West. Later, when the dark veil of medieval religious intolerance began to lift, Arabic texts and translations allowed Western European scholars to rediscover and reconstruct a vital legacy.

One of the earliest Arabian scholars of note, Abu al-Jahiz (776–868) reputedly wrote more than 200 works. His most famous, the *Book of Animals*, is an encyclopedia in seven large volumes, in which he discusses such diverse subjects as the social organization of ants and the effects of diet and climate on men and animals. There is a legend that, as a bibliophile to excess, he was accidentally crushed to death by a collapsing pile of books in his private library.

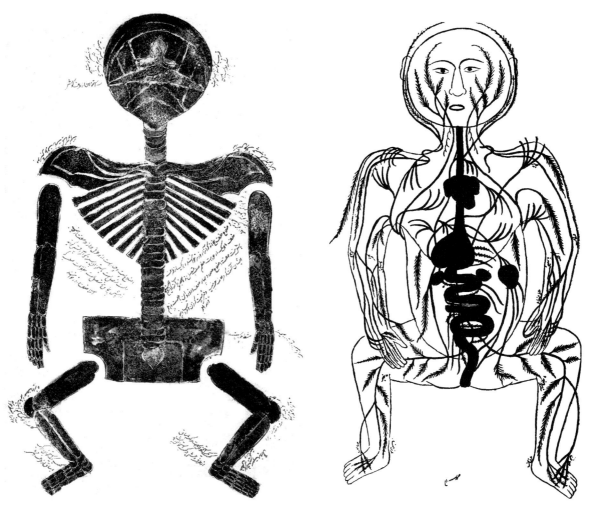

Fig. 8.1. Islam conserved and advanced Greek science, lost in the West. Persian human skeleton. Dated around 1200 AD (Choulant, 1920). Original in British Library, London.

Fig. 8.2. A representation of the venous and digestive systems, from a Persian manuscript dated about 1400. (From Choulant, 1920.)

Al-Razi (Rhazes) (869–925), of Baghdad, wrote over 100 works, and was regarded as the greatest of the Arabic-writing alchemists, and seems to have been the first to make the division of substances into animal, vegetable, and mineral (Singer, 1959). His encyclopedic *Al-Hawi* (*The Content of Medicine*) gathered into one huge book (one surviving Latin edition weighs 24 pounds) the medical knowledge of Greece and the Middle East, together with his own contributions (Castiglioni, 1958). It was translated into Latin by order of Charles I of Anjou, King of Sicily, by the Jewish physician Farragut, in 1279, and was repeatedly printed after 1488. *Al-Hawi* was known as *Liber Continens* in its Latin translation. Al-Razi also wrote well quoted books on diseases and pestilences.

Alhazen (Abu Ali Hasan Ibn al-Haitham, 965–1043) made many important new discoveries in optics. In his translated books on optics, and light *Opticæ Thesaurus* and *De Luce*, he gives the most accurate contemporary description of the anatomy of the eye and its nerve connections (Hatfield and Epstein, 1979). A diagrammatic illustration from Alhazan's book on optics is depicted in Fig. 8.3 showing the structure of the eye, and its connections via the optic chiasma to the brain.

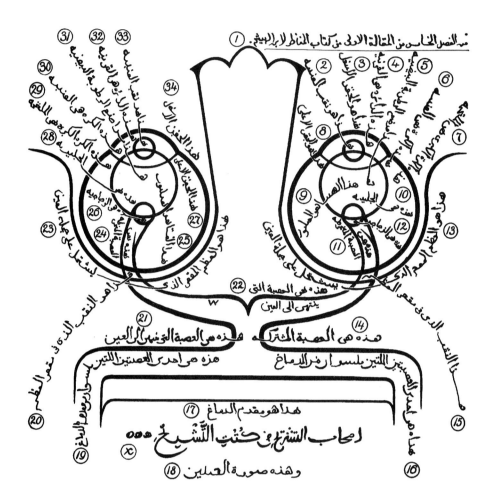

Fig. 8.3. Schematic eye showing details of eye and brain connections including the optic chiasma. From Alhazens (965–1039) book of optics. This may be a copy of a Greek original. (From Polyak, 1941.) Reproduced with permission from the University of Chicago Press.

Alhazen appears to have conducted experiments on the propagation of light and optic illusions and reflections, and ridiculed earlier theories of vision that the eye sends out visual rays to the object of the vision; if this theory were correct, he writes, then many observers would improve perception, since more pneuma would be released from the eye into the surrounding air to further illuminate the object. According to Alhazen, the rays originate in the object of vision and form an image on the anterior surface of the crystalline lens. Using the most recent anatomical knowledge, he adapted Ptolemy's visual pyramid geometry to explain this process. He maintained that the image is transmitted through the vitrous humor, and down the hollow optic nerves. In the optic chiasma, the separate transmissions from each eye join to form a single impression, which is received by the faculty of sense that completes the act of vision. But, in an important insight, he declares that this interpretation cannot be passive. The sense uses "reasoning" and "distinguishing" to make rapid visual judgements, so rapid that they are unperceived. This explains why we apprehend a circle as a circle, even when it is presented at different angles to the horizontal, and it is how we are able to determine

EVIDENCES OF THE FOUR PRIMARY INTEMPERAMENTS

Evidence.	Hot.	Cold.	Moist.*	Dry.
Morbid states to which there is a tendency	Inflammatory conditions becoming febrile. Loss of vigour.	Fevers related to the serous humour. Rheumatism. —	— Lassitude.	— —
Functional Power	Deficient energy.	Deficient digestive power.	Difficult digestion.	—
Subjective sensations.	Bitter taste in mouth. Excessive thirst. Sense of burning at cardiac orifice.	— Lack of desire for fluids. —	Mucoid salivation. — Sleepiness.	— — { Insomnia. { Wakefulness
Physical signs.	Pulse extremely quick and frequent; approaching the (weak) type met with in lassitude.	Flaccid joints.	Diarrhœa Swollen eyelids	Rough skin. Spare habit (acquired not inborn).
Foods and medicines.	Calefacients are all harmful. Infrigidants benefit.	Infrigidants are all harmful. Calefacients benefit.	Moist articles of diet are harmful.	Dry regimen harmful. Humectants benefit.†
Relation to weather (i.e., season).	Worse in summer.	Worse in winter.		Bad in autumn

Fig. 8.4. The famous Table of Temperaments from Avicenna's *Canon*. (Translation by Gruner, 1930.)

such features as distance, size, and shape. The concept of a sensory core that interprets visual stimuli was highly sophisticated, incorporating mathematical, anatomical, and physiopsychological components. Alhazen also provided a model for binocular vision, and explained the apparent increase in size of the sun and the moon when they are near the horizon.

The most renowned of the Arabian medical scholars is Avicenna (Abu Ali al-Hussein ibn Abdallah ibn Sina) (980–1037), known throughout the Middle Ages as "the Prince," the rival of Galen (Osler, 1921). He wrote the famous *al-Qanun*, or *Canon of Medicine*, which integrated all Greek and Moslem medical and biological knowledge. The *Canon*, an enormous undertaking, ran to more than a million words, the subject matter being arranged systematically in five large volumes. Volume I dealt with the principles and theories of medicine;

Volume II treated the simple medicaments; Volume III described diseases associated with parts of the body, such as intestinal diseases; Volume IV described diseases associated with the whole body, such as fevers; and Volume V dealt with the composition and preparation of drugs.

Avicenna's *Canon* not only reviewed all medical knowledge available from ancient and Muslim sources, but it was also enriched with important commentaries and new details. The temperaments, for example, were extended to encompass emotional aspects, mental capacity, moral attitudes, self-awareness, movements and dreams (Fig. 8.4), a forerunner of twentieth century psychoanalysis. The *Canon* was translated into Latin by Gerard of Cremona in the twelfth century, and became the standard textbook for medical education in Europe. Editions continued to be used down to the middle of the seventeenth century. Avicenna also wrote

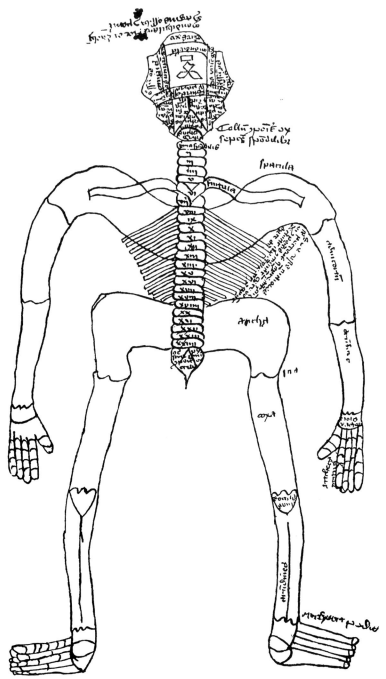

Fig. 8.5. Western medieval copy of Arabic original. Fourteenth century Latin illustration of human skeleton. (From Choulant, 1920.)

books on healing, as well as synthesis of all knowledge—abstract, practical, and theological.

Avicenna also made many original contributions, particularly in optics. He described the six extrinsic muscles of the eyeball, and provided new detailed descriptions of different parts of the eye, such as conjunctive sclera, cornea, choroid, iris, retina, layer lens, aqueous humor, optic nerve, and optic chiasma.

Ibn Rushd, or Averroes (1126–1198), born in Granada, Spain, was one of the most influential of medieval thinkers, and was reputedly the greatest of all Arabic commentators on Aristotle. He wrote extensively on medicine and particularly on optics. His most frequently quoted book, the *Colliget*, a seven-part summary of medicine, is more of a philosophical discussion on medicine than a textbook.

Ibn-el-Nafis (1208–1288), born near Damascus, wrote many books, including *Mujaz Al-Qanun*, a *Summary of the Canon*. Here he criticized what he saw as mistakes in Avicenna's book, and added new observations. He made particularly novel observations on the physiology of circulation, denying Galen's invisible pores in the intraventricular heart septum, and suggesting that blood flows from the heart through the lungs, where it mixes with air, then flows back to the heart (Bittar, 1956). In his own words, from the translation in Rothschuh (1973): "When the blood in the right chamber has been refined, it must reach the left one where the vital spirit originates. Now there is no communication between these two cavities, ...not even invisible pores, like Galen has postulated."

Ibn-el-Nafis then proposes that the blood in the right chamber flows to the lung, "...where it spreads across the pulmonary substance and mixes with the air, ...whence to the left cardiac chamber." This is the first recorded description of the pulmonary circulation.

As in medieval Europe, the bestiary was also popular in Arabic scientific literature, the most successful being the Ibn Buhtisu bestiary (Contadini, 1994), which provided moral lessons from the behavior of animals, such as the combativeness of the ram, the jealousy of the bull, and the thievery of the magpie. But it also contains quite practical information. Dried camel lung is a recommended ingredient in a concoction against asthma, and, "if a woman no longer a virgin eats the jugular vein of a sheep she becomes a virgin again" (Cantadini, 1994), with the hopeful comment, "this is proved!"

The influence of Arabian scholarship on the renaissance of Western science cannot be underestimated. The Western revival of learning in the eleventh to thirteenth centuries was, to a great extent, fueled by translations of Greek manuscripts from Arabic sources. An example of this dependency is shown by comparing the skeletons in Figs. 8.1 and 8.5. The Arabian depiction of a human skeleton (Fig. 8.1) dated around 1200 AD, is thought to be of ancient origin, perhaps Alexandrian (Choulant, 1920). It depicts the skeleton from a peculiar perspective, a back view, with the head strongly bent backwards, so that the upside-down head is facing outwards, with the chin uppermost. The fourteenth century Latin skeleton illustration (Fig. 8.5) is almost identical in its overall design: It is a back view, with the skull bent back so that the face, nose, and mouth are visible, and the highest feature is the chin.

Indeed, at this period, Arabic was the language of the European scholar, and, in translations of the high Middle Ages, the scientific nomenclature was predominantly in Arabic. However, in the early Renaissance, the Arabic medical terms were translated into their Greek and Latin equivalents. Nevertheless, an Arabic linguistic influence still lingers in science; for example, in mathematics, "algebra;" in chemistry, "alchemy," "alcohol," "alkali;" and in biology, "colon" and "cornea."

A good example of how established and reputable Islamic authorities were in Western medicine is illustrated in Chaucer's *Canterbury Tales*. In the prologue, the Doctor of Physic, "a verray parfit praktisour" impresses the pilgrims with his knowledge:

"wel knew he of the olde Esculapius
And Descorides and eek Rufus
Olde Ypocra, Haly and Galen,
Serapion, Razi and Avycen,
Averrios, Damascien and Constantyn,
Bernard and Gatseden and Gilbertyn."

This list includes four Arab physicians Haly (Ibn 'Isa), Razi (Rhazes), Avycen (Avicenna), and Averois (Averroes). Constantyn is Constantine the African, a monk at Monte Cassino, and a prolific and important translator of Arabic into Latin texts. But apparently he made one too many translations, because Chaucer later condemns Constantine as the "cursed monk, dam Constantyn," author of the book *De Coitu* which deals with aphrodisiacs and sexual intercourse.

As an interesting perspective, it has been suggested that the Arabs were not particularly interested in Greek or Roman poetry, plays, or history, so that when Greek knowledge first came to be transmitted to the West through translations of Islamic texts, it was at first limited to science and philosophy. The classical humanities were rediscovered for the most part directly from Greek texts later in the Renaissance. It has been argued that this breach was a source of the lingering intellectual separation between science and humanities. (Singer, 1959)

9 Scholasticism and Science

The late Middle Ages saw a revival of interest in science, as Islamic and Hebrew translations of Greek texts became available. This stimulated the rise of the universities and the development of scholasticism. This "science" was utilitarian: Astrology, alchemy, and magic were respectable disciplines. Complex intellectual constructs were built to account for nature, e.g., the macrocosm–microcosm, parallelism, the great chain of being, and the cell doctrine of the brain. Inspired by the rediscovery of Galen (via Avicenna) and the enormous richness of Greek biological concepts, there was a revival of human dissection. Initially, these served to demonstrate the truths of Galen.

The flood of new knowledge—much of it old knowledge, rediscovered—stimulated a great increase in intellectual activity in the late Middle Ages (Crombie, 1959). Twelfth century translations of Aristotle were especially influential, and the practice of Aristotelian logic revitalized man's belief in the rational faculty (Stiefel, 1977). Visions were no longer sufficient to reveal the true construct of the universe, and, although reason had been denigrated during the Dark Ages, it now reigned supreme. The existence of God could be proved by reason, and God had designed the world to exhibit, in the highest degree, a rational and beautiful order. For example, according to Adelard of Bath, the existence of grass was not only a wonderful manifestation of God's will, but it also had a reason: to provide

nourishment for sheep and cattle. To the smallest degree, the structure and function of the human body also illustrated the wisdom and goodness of God, and it was believed that all knowledge was ultimately divine, and that all learning was primarily a means of salvation (O'Boyle, 1992). Accordingly, the sublime task of the scholar was to reveal this rational order. At this period, the scholar would most likely also be a cleric. Indeed, in some universities, such as the University of Paris, new students were tonsured as a sign of their clerical status.

Perhaps the most important development that facilitated this intellectual ferment of scholasticism was the rise of the universities in the twelfth and thirteenth centuries. They provided a long-lasting, stable environment devoted to teaching and knowledge. At first, the universities were ecclesiastical establishments, authorized by papal bull and manned by theologian-scholars. In 1156, a medical faculty was established at the University of Bologna, and Padua University was founded in 1222. The medieval universities became extraordinarily popular, and spread throughout Europe. By the fifteenth century there were hundreds, including two in England and three in Scotland. During this period, there was a gradual shift away from religious, toward secular (earthly) knowledge.

A standardized, two-part curriculum was commonly adopted, based on the medieval seven arts. The first part, the trivium (trivial), contained three studies, grammar, rhetoric, and logic, which were thought to provide an indispensable base for a command of language and the necessary skills for logical argument. The second part, the quadrivium, completed the liberal education with arithmetic, geometry, astronomy-astrology, and the laws of music. The quadrivium provided the essentials for the study of natural science. It is important to note that this is essentially a secular curriculum in natural philosophy. The award of a degree required formal qualifications. For example, in the medieval University of Paris, after finishing his basic education, the candidate for the bachelor's degree had to put forward propositions over the 40 days of Lent, and defend them against opponents. After a further two or three years' study, he could defend his licentiate, and several years later become a doctor of the Sorbonne, the highest intellectual distinction.

The importance of the universities, for the future development of science, is that they rapidly matured into destinations for scholars, to take part in learning and teaching, and to transmit the great rediscovery of the classics. Because handwritten books were scarce, only the new universities could provide a forum for lectures, disputations and demonstrations. And, most importantly, they were stable, surviving century after century, keeping alive a tradition of scholarship. The new universities not only preserved the classical texts,

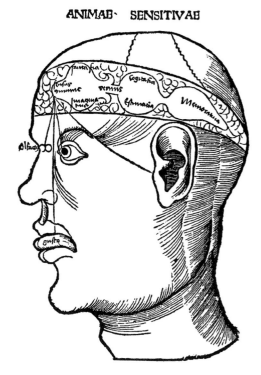

ANIMAE· SENSITIVAE

Fig. 9.1. Diagram of the three ventricles and their contents. The legends in the ventricles are: "sensus communis, fantasia/imaginativa, vermus, cogitativa/estimativa, memorativata." (From G. Reisch *Margarita philosophiae*, Freiburg, 1503.)

but now also studied them, even writing new treatises on the subjects.

By the thirteenth century, Aristotle's natural philosophy and Avicenna's *Canon* were fully established as the basis of the medical curriculum, and, during the second half of the thirteenth century, the scholastic method became the mode for the teaching and learning of medicine (O'Boyle, 1992). Here, truth was arrived at by critically comparing statements in authoritative texts, through a process of stylized rational analysis, and from elaborate arguments about textual interpretation. Scholastic philosophy was based on the distinction between the two forms of substance and accident. The form of substance makes the thing be what it is, and which, in its absence, would cease to be; an accidental form may be removed without the loss of the object. Lustre, a favorite example, is an acciden-

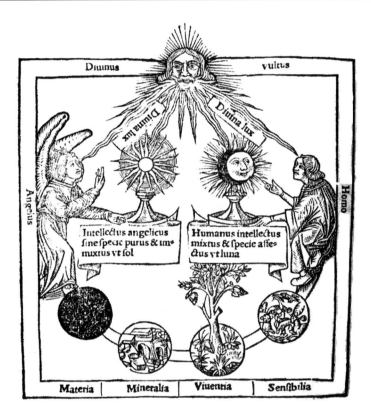

Fig. 9.2. Medieval concept of the great chain of being. Divine light illuminates both angels and man, who are linked, in a perfect circle, through the kingdoms of animals, plants, minerals, and unformed matter. (From *De Intellecto*, Charles Borillus, 1470–1550.)

tal form of gold. From this premise, it was reasoned that the true goal of the alchemists is to discover the substantial form of gold, which accounted for "goldness." But it was conceded that a substance may have any number of accidental forms, and scholastic disputes thrived over the identities of substance and accidents for particular objects.

In formal arguments, knowledge was arrived at through questions and answers formulated in Aristotelian terms. For example, the nature of medicine was a popular debate. Was medicine a "scientia" (i.e., true knowledge that could be derived from axiomatic principles) or an "ars" (i.e., a set of intellectual or mechanical skills to promote healing) (O'Boyle, 1992). There were arguments over the number and location of the internal senses. Following Galen, some scholars declared for four

internal senses; common sense (which received the five external senses), imagination, cognition, and memory; others maintained that there must be a fifth sense: fantasy. The senses were thought to reside in the three hollow chambers, or ventricles, of the brain, but the precise locations of the different senses was a never-ending dispute (Fig. 9.1).

As the process developed, intellectual schemes of great complexity were constructed. Parallelism was still greatly favored, the theme being that, from simple beginnings, more and more complex patterns are built. There was a natural hierarchical order in society, as well as in life (Fig. 9.2). In the heavens, God was at the head, followed by the archangels, the angels, and the lesser angels. The spiritual counterpart on Earth had the pope at the apex, the archbishops below, followed by bishops

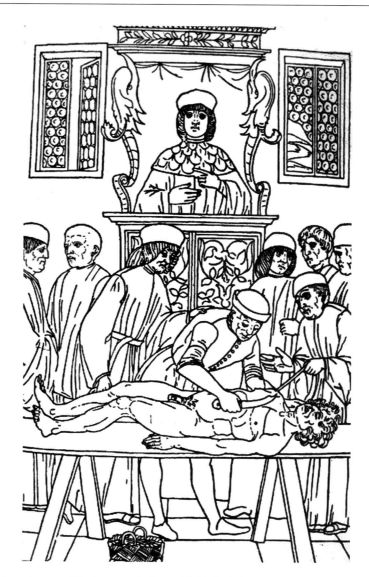

Fig. 9.3. The formal anatomy lecture. The lecturer sitting on his high chair reads out loud from a text derived from Galen's anatomy. The dissector, whose dress is distinguished by a row of buttons, is about to cut open the chest. His task is to illustrate the truths of the text. This drawing is from Johannes de Ketham's *Fasciculus medicinae*, 1493. It proved very popular, and was copied often, with variations.

and priests, with laymen at the bottom. The corresponding descending order of temporal society was the king or emperor, great nobles, barons, and, at the bottom, the serfs: God's design, with each in his proper place. The body was also arranged as a feudal system, the hot heart having the office of the king, and being like the sun; the cooling lung, the orator; and the brain being the chancellor and the seat of common sense that judges the senses (Singer, 1959).

The structure of the dispute was formal. "Dialectical teams" were assembled to present the arguments in favor of, or against, the contested opinion, in a series of debates; which were open to the general public, and would arrive at a final solution. Since the methods of investigation were equally applicable to resolving doctrinal disputes about the Christian religion, anyone who had mastered the scholastic method could move from one subject to another with little difficulty. Consequently, the lead-

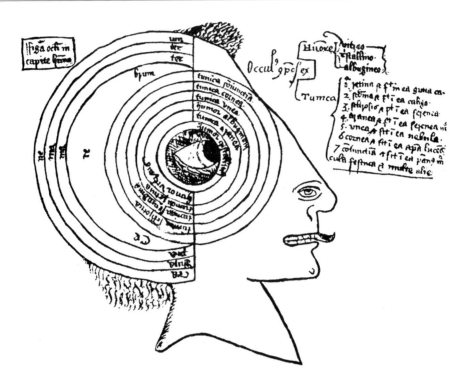

Fig. 9.4. The parts of the eye, from a late fourteenth century manuscript. Although it has a distinctly "surrealistic" perspective, the figure contains some accurate information. The labels on the right list the three humors and the seven tunics, including the retina (**1**) and the cornea (**6**). On the left, the cranium, dura mater, pia mater and the cerebellum are named. (Choulant, 1920.)

ing scientists and the leading theologians were often the same men.

The most illustrious of the Schoolmen, Albertus Magnus (1193–1280), was called "Doctor Universalis' and "the Divine Master." Although a Dominican theologian, he also wrote on natural science and medicine, and speculated much on the classification and localization of the internal senses. His two most famous pupils were St. Thomas Aquinas and Roger Bacon.

Roger Bacon (1214–1292), "the Wonderful Doctor," provides an example of the tortuous scholastic logic argument unhampered by observation or measurements. A Franciscan well-read in Latin and Greek, he accepted the new translations of Arabic medical and pharmaceutical works, but, on "logical" grounds, did not fully agree with Hippocrates and Galen. For example, he rejected Alhazan's model of vision caused by light emitted from the perceived object and accepted the Pla-

tonic idea of a beam shining from the eye, on the grounds that, by their nature, the emissions from objects are not suited to act immediately on sight, because of the "nobility of sight." Therefore, the perceived objects must be aided by and excited by the emission from the eye, which, altering and "ennobling" the medium, renders it commensurate with sight, and prepares for the approach of the emission from the visible object (paraphrased from Grüsser and Hagner, 1990).

In typical scholastic language, the third doctor, St. Thomas Aquinas (1225–1274), "the Angelic Doctor" explained, in his book, *Contra gentiles*, how, of all forms, the most perfect form must be the human soul. "Body and soul are not two actually existing substances, but out of the two of them is made one substance actually existing: for a man's body is not the same in actuality when the soul is present as when it is absent: it is the soul that gives actual being." Aquinas was famous for his "logi-

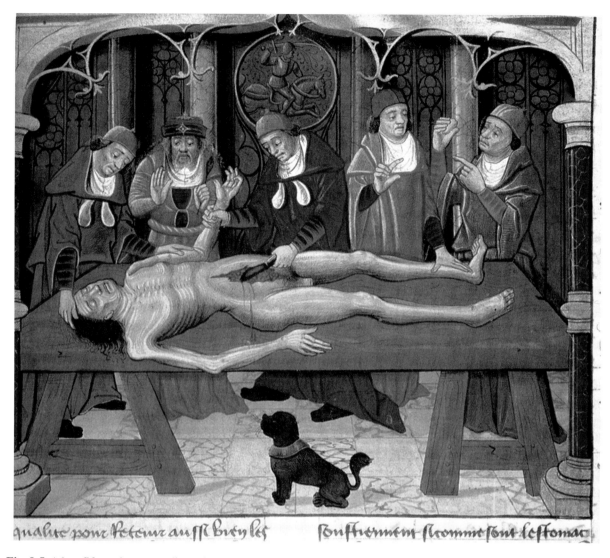

qualite pour Peteur auffi Brey les   souffrement ficomme font feftomac

Fig. 9.5. A late fifteenth century dissection. The bleeding indicates that the corpse is fresh. The inevitable accompanying dog looks on. (From *De propietatibus rerum* by Bartholomeus Angelicus, Bibliothèque National, Paris.) (*See* color plate appearing in the insert following p. 82.)

cal" proofs for the existence of God (Stiefel, 1977). The fifth proof invokes the harmony of nature. Everywhere we look there is harmony or accord. Fish need to swim, so they have fins and tails; dogs need to gnaw bones, so they have strong teeth. This accord is either by accident or design. Accident is inconceivable; therefore, this harmony in nature demonstrates there must be an supreme intelligence that organizes things. Design is still the basic argument of today's creationists.

The concept of "form" was also at the base of a long dispute on the role of nutrition in producing flesh. It was argued (and disputed) that two kinds of flesh could be distinguished. There is essential flesh, to which food contributes nothing. This flesh is related to true human nature, is drawn from the parents, and derived from Adam. Matthew 15:17 is a supporting authority here: "Whatsoever entereth in at the mouth goeth into the belly, and is cast out into the draught." The other flesh is the material flesh, which is affected by food, and is consumed (burned) to produce innate heat. This flesh decays on death. The distinction reflected a widespread belief that it is only the essential flesh, not made

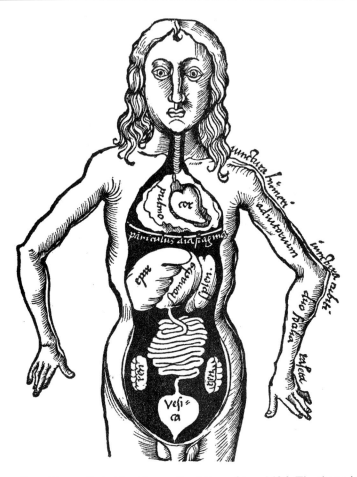

Fig. 9.6. Visceral anatomy from Gregor Reisch's *Margarita philosophica*, 1496. The thoracic and abdominal cavities have been dissected to show the most important organs in their natural settings. (From the National Library of Medicine.)

from food, that will constitute the resurrected person at the day of Last Judgement.

The age of scholasticism revived the art of rational, Aristotelian argument. Intellectually, the complex formal arguments of the scholars, in interpreting the divine will through scripture and nature, is paralleled in the intricate architecture of the great Gothic cathedrals, such as Notre Dame, with its elaborate interweaving motifs and magnificent flying buttresses. Corresponding intellectual development is illustrated in the elaboration in music, from Hildegard's plainchant to the extraordinary, convoluted, and beautiful polyphony (with as many as 15 independent lines sung simultaneously) of the thirteenth century. But, in the absence of new infor-mation, new facts, and new discoveries, the

arguments, no matter how clever, logical, and comprehensive, were ultimately sterile.

However, the rediscovery of Galen, through the treatises of Avicenna, was fomenting change. The ancient's comparatively sophisticated treatment of anatomy and physiology, with its richness in rational explanations about how the body was constructed and how it worked, showed up the utterly impoverished state of current knowledge. There was a revival of interest in human dissection and the use of the cadaver as a teaching tool. In 1231, the Holy Roman Emperor Fredrick II passed a decree that the human body was to be anatomized at least once in five years at the University of Salerno, and physicians and surgeons of the empire were required to be present. Unfortunately, there are no surviving records of these demonstrations. The most famous anatomist, Mondino (ca. 1270–1326), became professor at the University Bologna in 1290, and, in 1315, performed the first recorded public dissection of the human cadaver since Greek times. Criminals were normally the subjects for dissection, honest folks being terrified of having their body scattered when the final resurrection of the dead took place. His anatomy, *De anatomi* written in 1316, had comparatively brief and concise instructions. For example, the dissection begins with the opening of the abdomen by means of a long vertical cut, followed by a horizontal cut above the umbilicus. He first discusses the position and function of the stomach, then gives directions on its dissection. But, repeating old Galenic errors, he attributes five lobes to the human liver, and describes an imaginary duct that transmits black bile from the spleen to the stomach.

As illustrated in the 1493 publication of Johannes de Ketham's *Fasciculus medicinae* (Fig. 9.3) (a much copied figure that was also used as a frontispiece in later editions of Mondino's *De anatomi*), the dissection was essentially a formal demonstration. We see the professor, sitting in his high-backed chair, or cathedra, reads instructions and interpretations from a text of Galen. A junior colleague, the ostensor or demonstrator, points with a staff. There being a strong aversion to touching a dead body, this task was left to the menial barber, who cuts with a large knife and handles the organs. Students look on. Because there were no preservatives, dissections had to be quick. They were normally completed in four daily readings. Day one covered the bowels and intestines (the most quick to decay); day two, the thorax, lungs, and heart; day three, the brain. By day four, the stinking and de-

caying corpse encouraged a fairly perfunctory examination of the extremities and spinal cord.

Although Mondino's text was handwritten, and therefore laborious to copy, it was popular, and was adopted as a reading text by the Italian universities, and was used until the sixteenth century (making it perhaps the first required course book). It stimulated a series of colorful and imaginative anatomical texts. For example, an early fifteenth century rendering of the parts of the eye is shown in Fig. 9.4. Although it has an odd, almost surrealistic, perspective, it has some accurate information. Labels on the right list the three humors of the eye and its seven tunics, including the retina (1) and the cornea (6). On the left of the diagram is labeled, from left to right, the cranium, dura mater, pia mater and cerebellum. The start of an autopsy is shown in Fig. 9.5; the bleeding suggests a freshly dead corpse. Gregor Reisch's (ca. 1467–1525) famous text, *Margarita philosophica* (1496), contained an early illustration of visceral anatomy (Fig. 9.6). It shows, with reasonable verisimilitude, the position of the trachea, lung, heart, diaphragm, liver, stomach, kidneys, and intestine.

Scholasticism played an essential role in the emergence of science in the Renaissance. Late scholasticism was fueled by increasing translations of Aristotle (Rotschuh, 1973) and indeed it has been said that "The philosophy of Aquinas is Aristotle Christianized" (Enc. Brit., 11th ed.). Logic became the tool of scholastic disputations. The endeavor to support religious truths through reason (by argument) established the authority of reason and encouraged a belief in the rationality of the cosmos. This led to the questioning of traditional knowledge through systematic doubt (Stiefel, 1977) and became a foundation to interpret new discoveries, a fundamental intellectual prerequisite for the rebirth of science.

10 Renaissance I

The Birth of Science

The new birth of science in the fifteenth century was in essence a revolt against dogma as the source of all true knowledge. For the medieval monk or nun, truth was revealed through scripture and visions; for the scholastics, truth was achieved through scriptural authorities and Aristotelian logic. Now the Renaissance scholar sought truth through observation and reason. Authoritarian scholasticism was discarded. For science, this new approach was manifested in a profound dissatisfaction with authority, argument, and logic as the sole and sufficient means to deduce how nature is composed. Conclusions had to be tested by evidence; had to be demonstrated. The slogan could have been, "Do not tell me, but show me!"

In art, in a reaction against the medieval symbolic illustrations of religious ideas, the drive was to represent as closely as possible how objects really were. Naturalism, copying nature as faithfully as possible, was the theme. Leonardo da Vinci (1452–1519), was a genius of extraordinary originality and breadth, excelling not only in painting and sculpture, but also in mathematics and science, in which his investigations ranged from physics to biology.

Leonardo was not satisfied with just painting the surface of the body, but he wanted to reveal its internal structure and function. He made over 750 detailed sketches from first-hand observations of anatomical dissections, many of them more accurate than Vesalius. These include drawings of the structure of the heart, muscle, brain, and the fetus.

He also was an ingenious experimenter. To assist in discovering details, he cleverly injected colored solutions into veins, and, borrowing a technique from bronze casting, pored or injected molten wax into the ventricles of the brain (Fig. 10.1). It was much more important, of course, to describe the arrangement of the mind-containing ventricles than the structure of the surrounding tissue. He performed animal experiments, inserting, for example, a long needle through the chest wall of a pig and into its heart, in order to follow the heart's motion. As an engineer, he studied the antagonistic actions of muscles by making tape models.

Leonardo was particularly interested in the reproductive organs, until then a forbidden or hidden territory, and appeared to be both fascinated

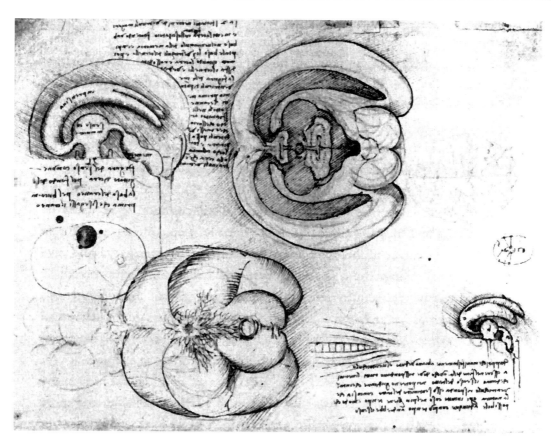

Fig. 10.1. Leonardo's sketch of the brain ventricles (ca.1506). With characteristic ingenuity, Leonardo injected molten wax into the ventricles to discover their shape when the wax hardened.

and repulsed by the sex act. One of his earliest anatomical drawings (ca. 1493) is of sexual intercourse (Fig. 10.2). However, Leonardo was also a man of his time. Following Plato's idea, that the soul-spirit-containing semen was secreted in the brain, he drew a hollow tube structure between the lower spinal cord and the penis for the passage of semen from the brain to the ejaculate.

Leonardo also indicated Galen's pores in his drawings of the septum of the heart. But, in accuracy, freshness, and vitality, his anatomical drawing far surpassed anything that had gone before. Ironically, his fame as an artist acted against the dissemination of his science. Most of Leonardo's biological notes on his ingenious experiments and observations were collected as curiosities by the English royal family and were held unpublished in their private Royal Windsor library, out of sight and unavailable to unprivileged view, until the beginning of the twentieth century, which was a deplorable loss.

But discarding the long tradition of scholasticism was not easy. Jacques Dubois, "Sylvius" (1478–1555), who was professor of anatomy at the University of Paris, followed the new fashion in teaching anatomy from human cadavers, and named many parts, including muscles, which, up until that time, had been merely numbered. However, although he made many anatomical discoveries, Sylvius remained a staunch member of the old school, regarding Galen as semidivine. He did not deny that there may be some differences in details between what is seen in dissections and Galen's descriptions, but the fault lay in the corpse being abnormal, or bodies may have changed since Galen's time.

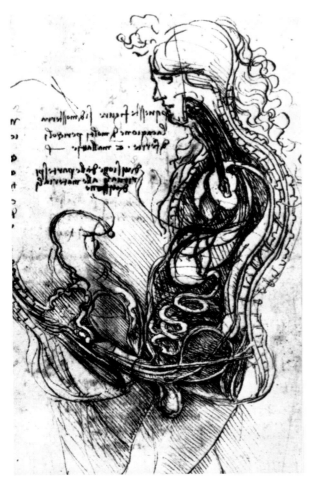

Fig. 10.2. Leonardo's drawing of human copulation. The figure has many erroneous "Galenisms." For example, he depicts a hollow tube structure between the lower spinal cord and the penis to allow for the passage of semen made in the brain to join the ejaculate.

Fig. 10.3. The magnificent frontispiece of Vesalius's *Fabrica* (1543). Eager to reveal anatomical errors, and to discover the truth, Vesalius stands next to the female body that he is about to dissect. The abdominal cavity has been opened. People of different classes throng about, arguing. The bearded man on the right sternly disapproves.

The major break with the past was provided by Andreas Vesalius of Brussels (1514–1564). In 1533, he went to the University of Paris, then medical center of world, to study anatomy under Sylvius. In 1537, he became the first chair of anatomy at Padua University, which was then one of the most prestigious universities, having obtained the complete Latin texts of Aristotle's biological writings. Boldly and courageously, Vesalius denounced the practice of using dissection as a mere illustration of Galen. He argued that structure must be discovered by direct examination, and he wanted to "put my own hands to the business." There was to

be no more magisterial promulgation from a high chair. Vesalius was an active participant in dissections, the new purpose of which was to test old, and find new, knowledge. His anatomical lectures were so popular that lessons had to be given in a special lecture hall (the aula), built to hold 500 people. Typically, the anatomy course lasted three weeks, with dissections each day on humans and animals. Drawings from the dissections were important tools in his teaching, and were made as accurate as could be.

In 1543, Vesalius's great book, *De humani corporis fabrica libri septum* (*Fabrica*) was published, which revolutionized anatomy and the study of the body. It was rich in understanding, and has beautiful illustrations, possibly by members of the famous artist Titian's workshop. The enthusiasm and appetite for the new knowledge is clearly conveyed in

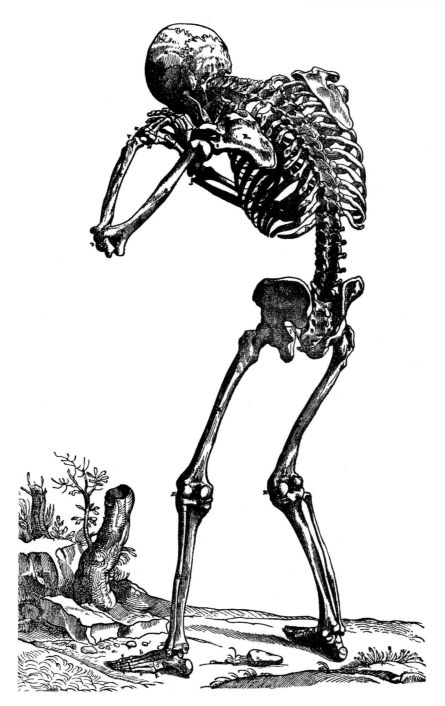

Fig. 10.4. Vesalius's weeping skeleton. Choulant (1920) comments that it "makes one think of the mourning apostles in a *'Burial of Christ'* by Titian."

the frontispiece, which shows an enthralled and arguing audience (Fig. 10.3). Vesalius himself does the dissection on a female corpse, "with his own hands" (Foster, 1901); an assistant below the table merely sharpens the knives. The audience, representing university, city, church, and nobility, is agitated, staring, or in dispute with each other. A bearded man on the right is looking away in deep disapproval.

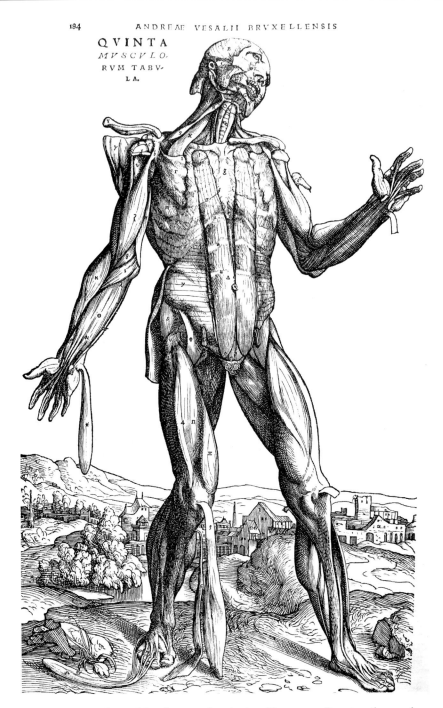

Fig. 10.5. Muscle Tabula from *Fabrica*. Muscles are often depicted in a state of contraction, and suggest movement and activity.

By an extraordinary coincidence, that same year, 1543, saw the publication of Copernicus's *On the Revolution of the Celestial Spheres*, which removed the earth from the center of the universe. According to Singer (1957), that was the year the Middle Ages died and the scientific renaissance was born.

Fabrica is composed of seven books. The first deals with bones and joints; the second, the most

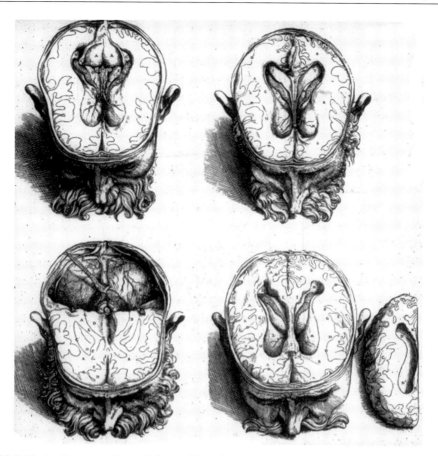

Fig. 10.6. Brain dissection from *Fabrica*. Note the emphasis is still on the hollow brain ventricles.

impressive, concerns muscles; the third, the vascular system; fourth, the nervous system; the fifth, the abdominal viscera; the sixth gives a somewhat superficial treatment of the heart and lungs; and the seventh is an outstanding description of brain anatomy. The power of the book lies in the large number of accurate and beautiful anatomical observations in highly dramatic illustrations. A skeleton is portrayed weeping (Fig. 10.4). Choulant (1920) comments that it "makes one think of the mourning apostles in a 'burial of Christ'" by Titian. Muscles are often depicted in a state of contraction, and suggest movement and activity (Fig. 10.5); the brain (Fig. 10.6), kidneys, and lungs are depicted in greater detail than ever before. Vesalius distinguishes between tendons and nerves, and discusses their function.

But, even beyond the illustrations, the tone of the book is important. Vesalius vigorously rejects Galen's authority, and does not hesitate to draw attention to his errors, listing about 200 in all. For example, most outrageously, he doubts the existence of Galen's pits in the septum, dividing the right and left ventricles.

However, Vesalius's enthusiasm for overthrowing old concepts sometimes led him to erroneous conclusions. A case in point concerns the function of the spleen. For Aristotle, the spleen was a companion to the liver, a "bastard" liver, and, like the liver, makes blood. Vesalius observed that the texture of the spleen was unlike that of the liver, and, from the aphorism, "Like makes like," concluded that the spleen does not make blood (Wear, 1977). In fact, although the spleen stores red blood cells, neither the liver nor the spleen makes blood.

The book finishes with an interesting chapter on physiological investigations: "On the dissection of living animals." Recalling experimental Galen,

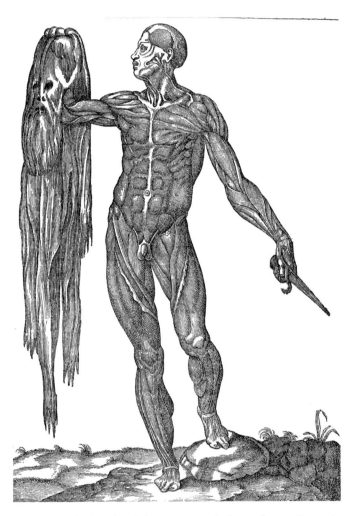

Fig. 10.7. A sixteenth century anatomical study of the outer muscle layer, from a front view of the body. This bizarre illustration of a man who has skinned himself is from *di Hamusco* (1556) by Juan Valverde (Choulant, 1920).

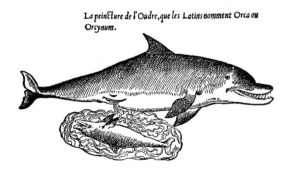

Fig. 10.8. There was a revival of interest (an Aristotelian interest) in discovering nature seen in such studies as a dolphin and its placenta. (From *Histoire naturelle des etranges poissons marins* [1551] by Pierre Belon.)

Vesalius showed that cutting the recurrent laryngeal nerves stopped vocalization. He found that excision of the spleen was survivable. Investigating muscle function, he found that cutting the muscle lengthwise did not stop contraction, but that cutting crosswise impeded contraction in proportion to the damage. From brain dissections of animals and humans he could see no differences that would indicate that the brain "function of animals should be treated otherwise than those of man; unless perchance… the mass of the brain attains its highest dimensions in man" (Foster, 1901).

Vesalius speculated rather conventionally on organ function. The kidney was fairly simple: It

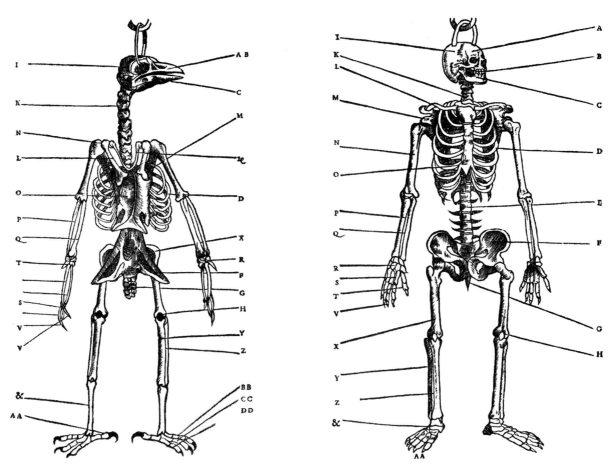

Fig. 10.9. Innovatively, the skeletons of a bird and man are compared bone by bone. (From Pierre Belon, *Histoire de la nature des oyseaux*, 1555.)

acted as a strainer of the arterial and venous blood, with which it was so richly supplied. The strained fluid, urine, gathered in the pelvis of the kidney, and was conducted by the ureters to the bladder. The function of the liver was thought to be well understood. The portal system carried nutritive material from the stomach and intestines to the liver, where a sort of fermentation took place, which made blood, in a process rather like changing grape juice into wine. The purer blood went, via the vena cava, to the heart, and the lighter impurity, the yellow bile, went to the gallbladder, and hence to the duodenum. However, he scoffed at the (Galenic) convention that a heavier impurity, the black bile, went to the spleen, from where it was transferred by a canal to

the stomach, to color the feces. He could find no such canal, and dismissed its existence.

Fabrica caused an initial storm of opposition. Vesalius's old teacher, Sylvius, felt betrayed, and wrote a furious and outraged rebuttal, *A Refutation of the Slanders of a Madman Against the Writings of Hippocrates and Galen*. He called Vesalius "Vesanus" (arse hole), "an ignorant slanderous liar, inexperienced in all things, ungrateful, and godless, a monster of ignorance who with his pestilent breath was attempting to poison all Europe, and whose errors were so numerous that merely listing them would be an endless task" (Ball, 1928).

But rejection was the exception, and *Fabrica* was enthusiastically accepted. It was reprinted many

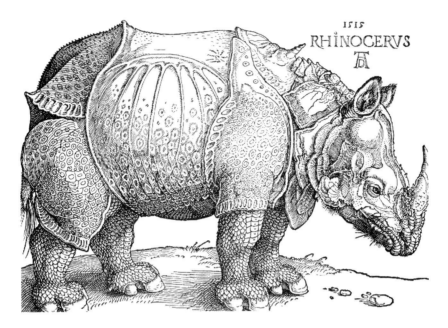

Fig. 10.10. Dürer's magnificently, but imaginatively, detailed rhinoceros (1515).

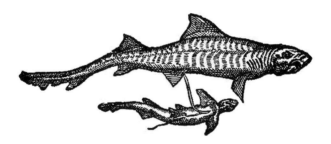

Fig. 10.11. A study of the dogfish. (From Guillaume Rondelet, *De piscibus marinis*, Lyons, 1554.)

times, and inspired a large number of similar books, many with the dramatic illustrations that had become fashionable. The "skinned man" by Juan Valverdi (1556) perhaps goes a bit too far, into the bizarre (Fig. 10.7): It depicts a man holding his own stripped skin to display the outer layer of muscle.

Excellent studies on comparative anatomy were also produced. In 1551, Pierre Belon produced a small book on the natural history of fishes, and another one on birds in 1555. He was the first to examine the placenta of the porpoise fetus, and to compare it to that of other mammals (Fig. 10.8).

Most originally, he made a comparative study of the skeletons of a bird and a man, illustrating them side by side and labeling the bone homologies (Fig. 10.9). The new naturalism in art is seen in Dürer's magnificent depiction of a rhinoceros (Fig. 10.10). Guillaume Rondelet (1509–1566) was said to have been so enthusiastic as an anatomist that he dissected the body of his own son (Singer, 1957). In his book on marine fishes (1555), he illustrated the placental dogfish that Aristotle had described, and that had been dismissed as error by later commentators (Fig. 10.11). Investigative dissection had been reborn.

Fig. 10.12. Old ideas and old creatures tenaciously held on. The dreadful lamia, with a beautiful woman's face and a serpent's tail, could entice young men to approach so that she might feed on their blood. (From Conrad Gesner's [1516–1565] *Historiae animalium*, translated and expanded by Edward Topsel [1572–1638] as *The History of Four Footed Beasts*, 1607.)

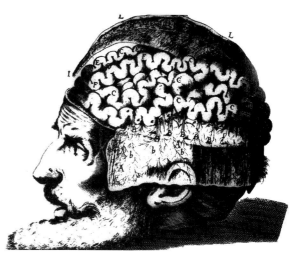

Fig. 10.13. A seventeenth-century illustration of the cerebrum, in which, still following the ancient Greeks, the convolutions are drawn to resemble intestinal coils (Caserio, 1627).

The switch from an ideological approach to nature, as seen in the bestiaries, to an attempt to describe how things actually are, is illustrated by Gesner's (1516–1565) *Historia animalium*. Gesner filled his massive 3500-paged *Historia* with the most accurate illustrations he could get. Instead of a mere catalog he arranged animals in "logical" categories, such as viviparous vs oviparous quadrupeds, and discussed them under such practical headings as names known by, habitat lived in, internal and external parts, use to humans, and so on. But the book, which was translated into English by Edward Topsel (1572–1638), still incorporated mythological unicorns and the lamia in its menagerie (Fig.10.12).

But what one sees is strongly influenced by what one believes, and old ideas die hard. Almost 100 years after *Fabrica*, the convolutions of the brain were still being drawn to resemble the coils of the small intestine, as described by the Greeks (Fig. 10.13; Casierio, 1627), and the medieval microcosm-macrocosm concept continued to flourish, as

gloriously depicted by Thomas Fludd (1574–1637) (Fig.10.14).

Some flavor of the slow and difficult transition to modern science is provided by Dr. Richard Burton's classic, *The Anatomy of Melancholy*, first published in 1621. This large treatise, of great erudition and learning, is a curious mixture of the medieval medical sciences and the embryonic modern approach. It is a fascinating book, to be savored and read in small pieces.

The humoral theory is eloquently presented. As quoted from the 1938 edition: "A Humor is a liquid or fluent part of the body... Blood is a hot, sweet, temperate, red humor, prepared in the meseraickik veins [in the liver]... Pituita or Phlegm, is a cold and moist humour begotten of the colder parts of chylus (the white juice coming out of meat digested in the stomack)... Choler is hot and dry, bitter, begotten of the hotter parts of the chylus and gathered into the gall. Melancholy, cold and dry, thick, black and sour, begotten of the most faeculent part of nourishment."

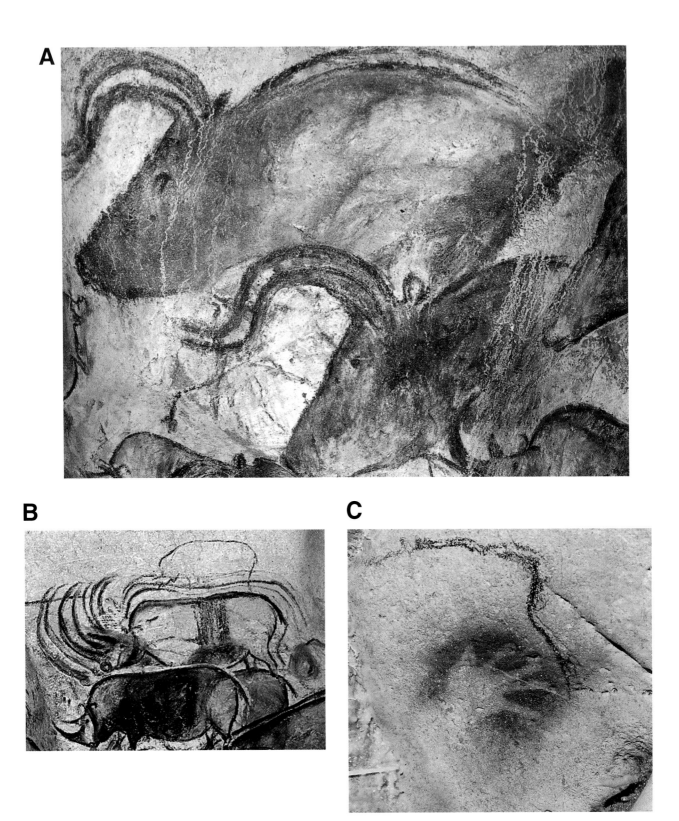

Plate 1 (Fig. 1.1 from Chapter 1). Graphic descriptions of nature are the most ancient human intellectual activity known. Illustrations from the Chauvet caves in Southern France, which contain the oldest known cave paintings, about 30,000 years old (Chauvet et al., 1996). Reproduced with permission of the French Ministry of Culture and Communication, Regional District for Cultural Affairs-Rhône-Alpes Region, Regional Department of Archeology. **(A)** Fine renderings of heads of the long-extinct auroch. **(B)** A herd of rhinoceroses. The decreasing size of the horns and the multiplication of the lines depicting the backs suggest perspective. **(C)** Stencil of the right hand of a 30,000 year-old-individual reveal a concept of self. The outline was made by pulverizing pigment on the hand flattened against the wall.

A

B

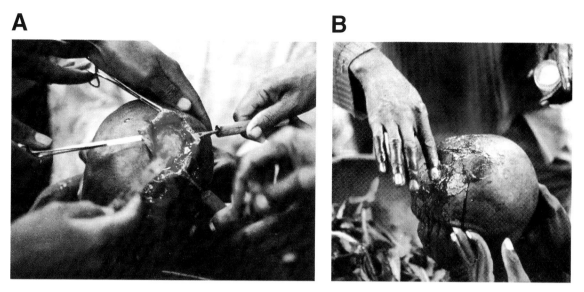

Plate 2 (Fig. 1.4 from Chapter 1). Trephination is still practiced today in parts of Africa, e.g., by the Kisii tribe in Kenya. (Reproduced with permission from Mueller and Fitch, 1994.) **(A)** A cross-shaped incision through the skin is made with a razor, the scalp flaps are reflected, and a hole is scraped through the skull bone with a hack saw. **(B)** After the surgery, the skin flaps are replaced and smeared with petroleum jelly.

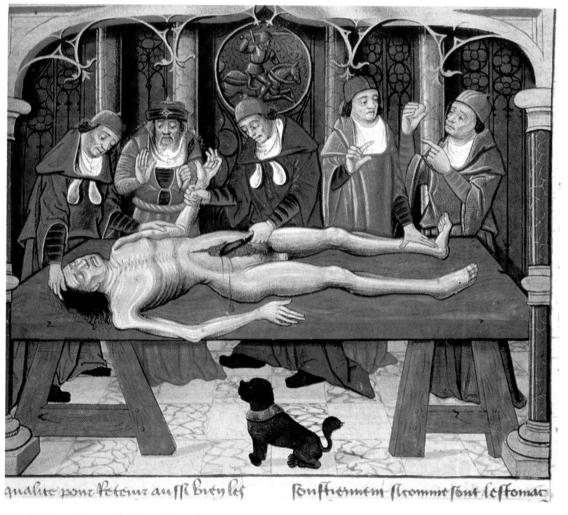

Plate 3 (Fig. 9.5 from Chapter 9). A late fifteenth century dissection. The bleeding indicates that the corpse is fresh. The inevitable accompanying dog looks on. (From *De propietietatibus rerum* by Bartholomeus Angelicus, Bibliothèque National, Paris.)

Plate 4 (Fig. 16.1 from Chapter 16). Hunter's transplant of a cock's spur into a cock's comb. (From Royal Society of Surgeons, England.)

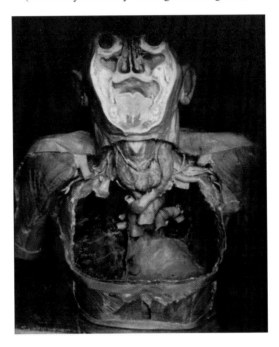

Plate 5 (Fig. 16.4 from Chapter 16). Anatomical display from the Hunterian museum, Glasgow University (S. Milton).

Plate 6 (Fig. 16.5 from Chapter 16). Mortsafes: Early nineteenth century graves armored against "resurrectionists." St. Mungo's Cathedral, Glasgow.

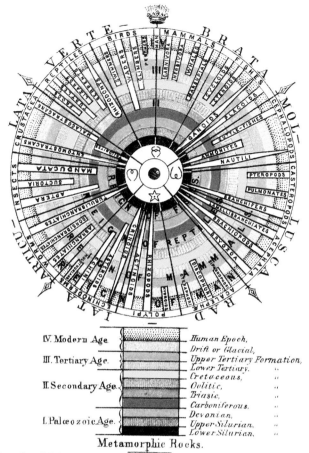

IV. Modern Age. — Human Epoch,
                    Drift or Glacial,
III. Tertiary Age. — Upper Tertiary Formation,
                     Lower Tertiary,   "
                     Cretaceous,   "
II. Secondary Age. — Oolitic,   "
                     Triasic,   "
                     Carboniferous,   "
                     Devonian,   "
I. Palæozoic Age. — Upper Silurian,   "
                    Lower Silurian,   "
Metamorphic Rocks.

CRUST OF THE EARTH AS RELATED TO ZOOLOGY.

Plate 7 (Fig. 18.1 from Chapter 18). Louis Agassiz's "physiological" interpretation of God's plan on the unfolding of life on earth for the preparation of man. (From Agassiz and Gould, 1851.)

Plate 8 (Fig. 19.5 from Chapter 19). Eugenics Society poster warning about the dangers of broadcasting bad seed.

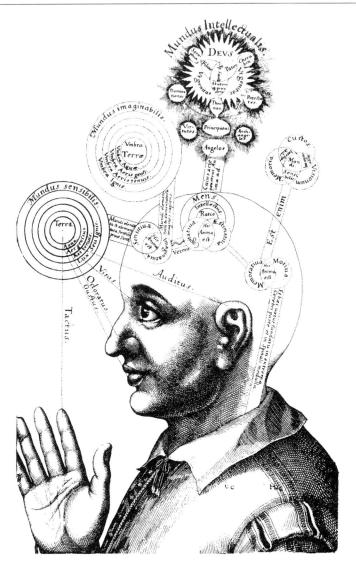

Fig. 10.14. The medieval world still rules in Fludd's world of the brain (1617). The three ventricles house sensation, intellect, and memory/motion in a complex interrelationship that links to the macrocosm, and to God and his angels.

Burton accounts for the function of the organs in dramatic scholastic terms: "The Heart is the seat and foundation of life... the Sun of our body... the King and sole commander of it... the seat and organ of all passions and affections.... The Lungs... are as an orator to the king... to express his thoughts by voice... also office to cool the Heart... .The Brain is the nobelest organ, the dwelling house of the soul" (Burton, 1938).

He describes that other realm, the world of spirits that saturated the universe, and one that we have quite forgotten: "Leo Suavius (Burton's reference)... will have the air to be as full of spirits as snow falling in the skies... and they may be seen by gazing steadfastly on the sun lighted by its brightest rays.... The air is not so full of flies in summer, as it is at all times of invisible devils. No place void but all full of devils... not so much as a hair breadth empty in heaven, earth or waters."

Burton provided rational explanations of mood: "Why students and lovers are often so melancholy and mad, the Philosophers of Coimbra assign this

Sprinto non spinto. More feard than hurt.

Fig. 10.15. The revolutionary new water closet had its hazards. (From *The Metamorphosis of Ajax* by the inventor, John Harrington, 1556.) Note fish in the water reservoir.

reason... because by vehement and continual meditation of that wherewith they are affected, they fetch up the spirits into their brain, and with the heat brought with them they incend (inflame) it beyond measure: and the cells of the inward senses dissolve their temperature, which being dissolved, they cannot perform their offices as they ought" — as good perhaps as many of the twentieth century universal explanations of human behavior, such as the unconscience, Freudian penis envy, the selfish gene, or the profit motive.

Burton also gives a quantitative estimate of the size of the universe: "Some of our mathematicians say that if a stone could fall from the starry heaven, or eighth sphere, and should pass every hour 100 miles, it would be 65 years or more, before it would

come to ground ... [The] distance heaven to earth [is] 170,000,803 miles."

And, finally, here is Burton's model of robust, self-serving behaviorism: "Cato would make his servants children suck on his wife's breasts, because by that means they would love him the better"(Burton, 1938).

The Renaissance was clearly a time of new, fresh thinking, and new approaches in science. Intellectual underpinnings for this new science were provided by Francis Bacon (1561–1626), Lord Chancellor to James VI of Scotland, simultaneously James I of England. Bacon's most famous book, *Novum organum* (1620), a direct challenge to Aristotle's *Organum*, is a collection of aphorisms to illustrate Bacon's new method. It is a direct attack on Aristotelian philosophy, which he refers to as idolatry.

Bacon argued that causes are hidden among the facts, and that the purpose of science is to uncover these causes. This is to be done by his method of induction, which required, first, the collection of all details associated with the phenomenon under investigation, then tabulation and classification of the results. This process of classification enabled one to discover those elements that were common and essential to the phenomenon under investigation, and justified the dismissal of those that were accidental.

He argued that, because the "essences" were all linked, in practice it is sufficient to know only a few phenomena well, because this would allow one to determine the key essences and know all. But he severely admonished that data must be taken without any presupposition, that it must be gathered blindly; otherwise, bias would influence what is recorded. This concept of the scientist as a disinterested collector of information was very influential for the future philosophy of science, but has often been ignored in practice.

For Bacon, the subtlety of nature would always be much greater than our understanding. Since sight cannot "penetrate into the inner and further recesses of nature," he thought that his inductive method was necessary to overcome this human perceptual limitation to inquiry. Although highly regarded in the history of science, as Wilson (1995) observes, Bacon's method of cataloging similarities is ultimately superficial. It cannot, for example, distinguish between the heat of pepper, the heat of burning, or the heat of friction.

Paracelsus (Philippus Aureolus Theophrastus Bombastus von Hohenheim) (1493–1541) was an important influence for change during this period. He was one of the most remarkable characters of the sixteenth century. Rude, rough-mannered, intemperate, and immensely conceited—he named himself "Paracelcus," i.e., "beyond Celsus" the celebrated Roman authority—he was an intemperate and passionate advocate against the established medieval practice of rote learning from authority. Upon his appointment to Professor of Medicine at Basel in 1527, Paracelsus publicly burned Avicenna's *Canon* and the works of Galen—in a gesture imitating Martin Luther's burning of the papal bull in 1520. He unceasingly attacked the medical establishment, writing in popular German, not scholarly Latin. "Medical faculty are vain glorious baboons in all their wealth and pomp, and there is no more in them than in a worm-eaten coffin… all they can do is gaze at piss" (Pachter, 1951). Naturally, this was not well-appreciated by his professional peers.

Paracelsus scornfully rejected the four Greek humors, but substituted three chemicals, mercury (volatility), sulfur (combustibility), and salt (the residue) as the basis of life. All physiological processes are governed by spiritual forces, the archaei, which bring about change in the body. The chief archeus is the soul, and death is its loss. He basically was providing a chemical explanation of life, and he urged that the goal of alchemy should be diverted from making gold to discovering life-healing drugs.

Paracelsus was a passionate champion of folk medicine. For example, it was thought, at that time, as of old, that pus formation was a normal part of the healing process: it was even called "laudable puss". Pus represented bringing out the badness in blood. So understood, standard medical practice encouraged pus by dressing wounds with special ointments or concoctions of cow dung, viper fat, feathers, and the like. Paracelcus scorned this "idiocy," and advocated in its place the folk medicine, "weapon ointment" or "weapon-salve." Weapon salve was an ointment made out of skull moss, flesh from the body of a hanged man (the fresher the better), and the patient's blood. Healing is effected when the balm is applied, not to the wound, but to the sword, or weapon, that made it. Ironically, it is almost certain that his nonintervention treatment saved many more lives than the medically approved dung therapy. One could imagine that wound salve would have been recommended by any contemporary epidemiological study, but correlation does not show cause.

As an aside, it should be remembered that contemporary ideas of hygiene are very recent. The early Christain church denounced the Roman baths as fleshly luxury, and it was regarded as saintly to abstain from washing. Filthiness was close to godliness. It was said that St. Anthony was never guilty of washing his feet, and, as the body of murdered St. Thomas a Becket cooled, contemporary accounts wondered at the vermin, which were living in his robes when they started to crawl out. "The vermin boiled over like water in a simmering cauldron, and the onlookers burst into alternate weeping and laughter" (Zinsser, 1934). It was not until 1556 that the water closet was first introduced to England by Harrington (Fig. 10.15), and, even as late as the eighteenth century, vermin such as lice, fleas, and bugs were thought to be natural products of putrefaction and decay of human skin and sweat, and normal inhabitants of the human body.

In his book, *A Treatise Concerning the Nature of Things*, Paracelsus reveals a dreaded secret, that an artificial man could be produced from putrefaction:

"Let the sperm of a man by itself be putrified in a gourd glass, sealed up with the highest degree of putrefaction in horse dung, for the space of forty days or so or until it begin to be alive, move and stir, which may easily be seen. At this time it will be something like a man, yet transparent without a body, Now after this if…[it] be fed with the arcanum (a secret essence) of man's blood and for the space of forty weeks kept in a constant heat of horse dung it will become a true living infant, which is born of a woman, but it will be far less."

Clearly, Paracelsus is not important as a scientific innovator, but his iconoclastic and protomaterialistic approach was new. His vigorous and intemperate break with the past had a marked influence on his contemporaries, and his conception of life phenomena as fundamentally a chemical process encouraged a new perspective for material explanations of life.

Paracelsus was however, an exception in his time. Jean Fernel (1497–1558), physician-in-chief to Henry II of France, was a much more respectable contemporary of Paracelsus ("deliberate and sober," [Hall, 1969a]). Fernel's famous text on anatomy, *Natural Part of Medicine* (1542), was published just one year before *Fabrica*. But he is old-fashioned in approach, still immersed in medieval ideas, with their elaborate schemes. For example, in his book, *On the Hidden Causes of Things* (1548), he distinguished between the simple parts of the body (bone, cartilage, ligament tendon, nerve, artery, vein, flesh, and skin) and the not-simple parts (such as blood, milk, hair, and nails). The complex parts are composed of the simple parts. The simple parts, in turn, are composed of the four elements mixed in proportions to produce a particular temperament, rather like, in music, harmony is a proportion among tones. Proper or "just" mixtures produce "moderate and concordant mediocrity" (Hall, 1969a). Here, he essentially follows Galen's 1400-year-old spirit physiology, with minor modifications. Natural spirit, made in the liver, from an especially pure part of the nutriment, passes to the right ventricle of the heart. It then moves through pores to the left ventricle, where, with the assistance of air coming from the lung, it is converted into vital spirit. The vital spirit moves up the arteries to the brain, where, in the rete mirable, it meets air coming from the nostrils, and enters directly into the brain ventricles, where it is transformed into a spirit, which allows movement and feeling and rational functions. The scientific rebirth was a curious and contradictory mixture of the old and new.

11 Renaissance II

The Birth of Experimental Biology

Ancient authority was no longer the arbiter of truth; explanations of how the body works had to be tested by observation, experiment, and calculation.

The initial steps in the renaissance of science in the Renaissance, dealt with in the last chapter, rejected classical authority, which instructed how nature really was, as against how it appeared. Dogma was replaced by a reliance on first-hand observation mode, as accurately as possible, to reveal new truths. The next step was the construction of demonstrations, set up to illustrate how a feature of the body worked. The basic premise was: A proposed function, if true, will result in a certain consequence, which is tested by experimental demonstration. This new method of experimental manipulation was the breakthrough to modern science. It is exemplified in the discovery of the circulation of the blood.

According to Galen, the liver made blood, and the veins, which originated in the liver, transported this nutritious fluid to the rest of the body, where it was consumed. Vital spirit was made in the left chamber of the heart, and moved along the arteries,

which originated in the heart. By this means, the arteries provided essential heat and vitality to the body. The arterial and venous systems communicated through tiny holes in the ventricular walls of the heart, which allowed some more subtle blood to pass into the left ventricle, where it was converted into this spirit. Although this scheme was impressive, in the sense that it appeared to explain all that was known, and was universally accepted, there were occasional critics. Galen's confusing explanation of the blood–air flow between the heart and the lungs was found particularly unsatisfactory by these "heretics," some of whom proposed that blood, not air, might flow to and from the lungs.

The most important early critic was the thirteenth century Arabian philosopher, Ibn el-Nafis, who, as we have mentioned before, denied the existence of interventricular pores, and described a pulmonary circulation that linked arteries and veins. But, unfortunately, this work appears to have been unknown, or at least unmentioned, in the West until 1922, when it was discovered in the Prussian state library (Singer, 1957).

The next potentially significant challenge to the Galenic system came 300 years later, when the scientist, medical doctor, and theologian, Michael Servetus (1511–1553), declared, in his book, *Christianismi restitutio* (1546), that "The vital spirit had its origin in the left ventricle of the heart and the lungs greatly assist in its generation" (Foster, 1901). He went on to state that the blood is "elaborated by the lungs, becomes reddish yellow and is poured from the pulmonary artery to the pulmonary vein," and, more significantly, "air mixed with blood is sent from the lung to the heart." He boldly declares that this is "a truth which was unknown to Galen." He did not explain how he came to these radical conclusions, and some have suggested that he knew of Ibn el-Nafis' work, but he appears to have been first in the West to so openly question Galen's explanations (Cournand, 1982).

Servetus was a biologist of outstanding promise. He had met Vesalius, and had been praised for his skill in dissection. But it was as a theologian that he attracted fatal attention. He was arrested by the Inquisition in Catholic Vienna, and charged with heresy. He escaped and fled to Protestant Geneva, only to be arrested again, and tried as a heretic. Under the orders of Calvin, he was condemned and burned at the stake, together with his books. Only a few copies of *Christianismi Restitutio* were missed (Fulton, 1953).

Realdus Columbus (1516–1559), a pupil of Vesalius, had also indicated some appreciation of a pulmonary circulation. His anatomical work, published in 1559, contained a chapter on vivisection, with observations on the movement of blood. He described a venous system coming from the heart, which "convayes blood to the lungs where it mixes with air and is conveyed back to the heart." "This," he rather ominously adds, "no one has hitherto observed and can be verified by (unspecified) experimental subjects whether alive or dead" (Cournand, 1982). He also noted that the pulsa-

ΕΧΕΡΟΙΤΑΤΙΟ
ANATOMICA DE
MOTV CORDIS ET SAN-
GVINIS IN ANIMALI-
BVS,
GVILIELMI HARVEI ANGLI,
Medici Regii, & Professoris Anatomiæ in Col-
legio Medicorum Londinensi.

FRANCOFVRTI,
Sumptibus GVILIELMI FITZERI.
ANNO M. DC. XXVIII.

Fig. 11.1. Frontispiece of Harvey's *De motu cordis* (1628).

tions of the brain were synchronous with the pulse of the arteries.

Although there appeared to be some scattered concept of the "lesser" pulmonary circulation, it required the genius of William Harvey (1578–1657) to discover, demonstrate, and prove the "greater circulation," and to radically advance the experimental approach in physiology. Harvey ("a little choleric man," according to the essayist Aubery), physician to Charles I, was responsible for substantial advances in biological concepts, and, indeed, has been called a founder of mechanistic biology (Hall, 1969a). As such, he deserves some special attention. But his enduring fame rests on his earlier revolutionary work, *Exercitatio anatomica de motu cordis et sanguinis in animalibus, (On the Motion of the Heart and Blood*

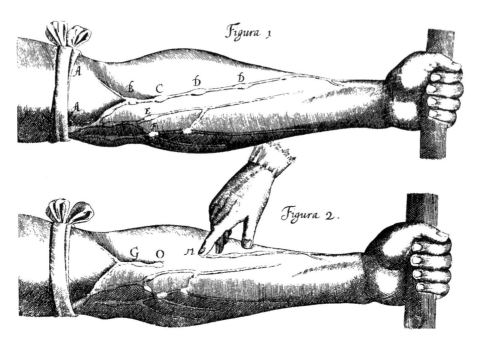

Fig. 11.2. Harvey's demonstration in *De motu cordis* of the action of the venous valves, which only allows blood to flow to the heart. Tying a light tourniquet at **A** causes, "especially in laborers," certain knots or elevations to be seen in the veins (**B–F**). Pressing one of these valves with a fingertip, and with another finger pushing the blood upwards/towards the heart, you will see this part of the vein (**O, M**) stays empty, and that the blood cannot flow back.

in Animals, 1628) (Fig. 11.1). This classic is a closely argued exposition of only 72 pages, almost just a tract, consisting of two major parts; the first concerns the function of the heart; the second, the circulation of the blood.

According to the old theory, the dilation of the heart was its most important function, in which acting like a bellows, it sucked blood from the veins to the heart. But this motion was not caused by any active participation of the heart, since it was believed that the walls of the heart were not muscular. By close and careful examination of the structure of the heart, Harvey found that the heart is indeed muscular, and he was able to show that the valves were arranged like "little gates," to guide blood flow in fixed directions through the heart.

Investigating the movement of the heart, Harvey found that the heart beat of mammals was too rapid to discern any detail, so he looked at the slow hearts of fish and snakes, as well as those of dying dogs and pigs. He carried out a series of beautiful experi-

ments on a variety of animals to determined movement of the blood, e.g., "If a live snake be cut open, the heart may be seen quietly and distinctly beating for more that an hour, moving like a worm and propelling blood. When it contracts longitudinally it becomes pale in systole, the reverse in diastole." He placed a clamp (or ligature) on the aorta, and found that the heart became swollen "like to burst." Placing a clamp on the vena cava caused the heart to become pale and to diminish in size.

These and many other experiments convinced him that the active movement of the heart is contraction, decreasing the volume of the chambers and driving the blood out. His conclusion, which was compelling to any unbiased observer, was that the heart receives blood from the vena cava, and its regular contraction drives the blood forward. Further, and against all established knowledge, he declared that the pulse is not an active expansion of blood vessels drawing blood out of the heart, but that it was a passive swelling caused by the propelled blood. Contra Galen, the beating of the heart is the cause of the pulse, not the other way round. He also established that the heart contracted from top to bottom, not right to left, as the Galenists believed.

Harvey's first major conclusion was that that blood was being pumped from the heart in two major streams, one to the body, the other to the lung. He vigorously denied the existence of the interventricular pores. How then did blood get from the right side of the heart to the left? His bold and revolutionary theory was that the blood must move in a circle through the lungs and back to the heart, then through the body back to the heart: the lesser pulmonary circulation and the greater body circulation. This he proved in the most original manner, by experiment and calculation. For example, in a series of experiments using ligatures, he was able to convincingly demonstrate that the blood flow in the veins was only in one direction, *toward* the heart (Fig. 11.2). He estimated the amount of blood pumped by the heart by such methods as severing a sheep's artery, and collecting and weighing the blood ejected in a specific time. He then made an extraordinarily original calculation:

"From experiment, I have found that the quantity of blood which the left ventricle of the heart will contain when distended, is, upwards of two ounces. Let us assume further how much less the heart will project into the aorta upon each contraction as approaching that the fourth, or fifth, or sixth, or even but the eighth part of its charge. This would give half an ounce ... of blood as propelled by the heart at each pulse into the aorta. Now, in the course of half an hour the heart will have made more than one thousand beats (lowest 33 beats/min), in some as many as two three, and even four thousand. Multiplying the number of ounces propelled by the number of pulses, we shall have either one thousand half ounces in 1/2 hr sent from this organ into the artery– a larger quantity than is contained in the whole body!" (Willis, 1847.) [Note: a 140 lb

person has about 8% blood = 11.2 lb blood = 180 oz!] The actual average is 72 beats/min = 4320b/hr × 1/2 oz = 2160 oz/hr = wt of body! [140 × 16 = 2240 oz.]

Harvey's grand conclusion, in one glorious sentence of 167 words:

"Since all things, both argument and ocular demonstration, shew that the blood passes through the lungs, and heart by the action of the ventricles, and is sent for distribution to all parts of the body, where it makes its way into the veins and pores of the flesh, and then flows by the veins from the circumference on every side to the center, from the lesser to the greater veins, and is by them finally discharged into the vena cava and right auricle of the heart, and this in such a quantity or in such a flux and reflux thither by the arteries, hither by the veins, as cannot possibly be supplied by the ingesta, and is much greater than can be required for mere purposes of nutrition; it is absolutely necessary to conclude that the blood in the animal body is impelled in a circle, and is in a state of ceaseless motion; that this is the act or function which the heart performs by means of its pulse; and that it is the sole and only end of the motion and contraction of the heart." (Translation by Willis, 1847.)

Harvey knew that he could not explain how blood makes the transfer through the tissues from arteries to veins, but he said that he simply did not know the answer, and did not want to conjecture. The solution had to wait until the invention of the microscope.

He fully recognized that he was overthrowing a deeply held doctrine: "I fear danger from the malice of a few." But, apart from "a few," his concept of circulation was widely accepted. James Pimerose (1592–1654), for example, denied all of Harvey's conclusions, especially his assertion that the interventricular pores did not exist. The pores are not to be found in dissection, he asserted, because they open and close as the heart dilates and contracts, and are permanently closed up, and thus not visible, after death. Caspar Hoffman (1572–1648) scoffed at Harvey's calculations, saying that the amount of blood passing through the heart is "a fact which cannot be investigated, a thing which is incalculable. Inexplicable, unknowable" (Whitteridge, 1971).

A new and troubling question arose. If the venous and arterial bloods were the same, why did they look so different? Harvey provided a mechanical explanation. The color difference came about because the blood then emerged through a narrow opening (artery) was the thinner and lighter part, while blood from the larger diameter vein was thicker and darker. Blood is an accidental property caused by straining (Frank, 1980). He showed that arterial blood was not rarefied blood, by a simple demonstration. Cutting an artery, he collected fresh arterial blood in a bowl. On cooling, it did not, as the Galenists would have predicted, "return to its original quantity of a few drops" (Willis, 1847).

Harvey was hesitant to speculate on the purpose of the circulation of the blood. In his opinion, there had been enough "idle speculation" and fanciful explanations by scholastics. But he was not free of old ideas: "The blood in the extremities, thickens from the cold and loses its spirit, as in death. Thus it must come back to its source and origin to take up heat or spirit or whatever else it needs to be refreshed." If circulation has a function, it is probably to convey warmth and nourishment from the heart (the center of innate fire) to the cold extremities, where the blood is cooled and thickened and taken back to the heart. More abstractly, he was impressed that blood circulation was a cycle and the cycle is the most perfect form.

Note that the movement of blood had been discussed and argued about for more that 2000 years, then it was solved at one blow. But, in principle, Harvey's approach could have been done at any time: It was not dependant on any new technology. Why not then? Why now? And indeed Harvey's

theory has been a *cause celebre* in tortuous contemporary debates on the philosophy and nature of science. In the end, it perhaps succeeded just because it provided the most natural and plausible explanation, using rational theory and experimental evidence in a new and convincing way (Mowry, 1985).

Harvey was interested in the broad sweep of life, and made extensive comparative studies on a wide range of animals, including toads, snakes, frogs, and deer. He was particularly intrigued by the ancient problems of reproduction and hereditary. He was an influential advocate for epigenisis, the idea that, during gestation, the embryo undergoes a progressive development from the undifferentiated seed to the complex newborn, and, in his book on embryology, *De generatione animalium* (1651), which presents his detailed research on the development of the chick embryo and fetal deer, he reports "that the first rudiment of the body is a mere homogenous and pulp jelly not unlike a concrete mass of spermatic fluid." He compared the developing fetus to a house or ship, whose frame is laid down first, and noted that, in the early stage of development, he could not distinguish between dog, snake and human embryos (Wilson 1995). Harvey made many further studies on the function of the lungs and the brain, on the senses and motion, and he was particularly interested in insects. But his unpublished notes were destroyed in the English Civil War (Frank, 1980). The loss was lamented by the celebrated contemporary poet Abraham Cowley; "Oh cruel loss … And ten times easier it is to rebuild Paul's (St. Paul's Cathedral in London), than any work of his."

At the time that Harvey was investigating the function of the blood vessels, another class of vessels was discovered: the lymphatics. In 1627, one year before Harvey's *De motu cordis,* a book by Gaspare Aselli (1581–1625) was published, called *De lactibus sive lacteis venis.* Aselli, professor of anatomy at Padua, described finding many white cords running over the intestine and mesentery of a recently fed dog. At first, he thought that he was dealing with nerve filaments, but, when he pricked them, he noted "a white liquid like milk or cream forthwith gush out ... . Seeing this I could hardly restrain my delight" (Foster, 1901). He thought he had at last discovered the route for the transfer of the famous chyle from the gut to the liver. He confirmed his observation by careful experiment, and named the new vessels "*venae albea et lacteae.*" Later, the Danish physician Thomas Bartholin (1616–1680) described this system in detail, and showed that these "chyle vessels" did not in fact connect with the liver. Bartholin was extremely pleased with his discovery, and concluded that food is not converted into blood in the liver as had been held from time immemorial. He triumphantly bragged that he had removed the liver "from its throne" (Foster, 1901).

12 The New Physiology

CONTENTS
IATROCHEMISTRY
IATROPHYSICS

The pace quickened with the search for the most scientifically comprehensive and all-encompassing explanations of life. Inspired by the new discoveries in chemistry and physics, two opposing schools arose to explain biological function, the iatrochemists (all is chemistry), and the iatrophysicists (all is physics).

The seventeenth century saw the founding, throughout Europe, of dedicated societies in which gentlemen members could discuss, exchange views, and publish letters on the latest developments in science. The Preussische Akademie der Wissenschaft was established 1670 in Berlin by Leibnitz; the Académie des Sciences was founded in Paris in 1666 by Louis XIV; and the Royal Society of London, which had existed as an informal group since 1645, received its Royal Charter in 1662. The motivation for these gatherings was a desire to see the new discoveries for oneself. The meetings were enlivened with demonstrations of the latest innovations and inventions, and the display of new measuring machines and astonishing biological specimens and geological samples (Fig. 12.1). Although wonder was excited, the elevated mission of these societies was to reveal the most scientifically comprehensive

and universal explanations of life. They provided a platform for the exposure of new hypotheses. Accounts of presentations were recorded in society proceedings, which also carried letters and commentaries from corresponding members, allowing distant "savants" to take part in deliberations and discussions.

In the nascent biological sciences, two rival life-explaining schools developed, the iatrochemists, who looked at chemistry to provide the answers, and the iatrophysicists, who advocated mechanical constructions. The temptation was to explain all on the basis of one simple cause.

IATROCHEMISTRY

The iatrochemists were inspired by alchemy, which, during the sixteenth and seventeenth centuries, had become sufficiently sophisticated in techniques to change natural materials into "something

93

Fig. 12.1. Seventeenth century biology in France. The inner organs of a fox are being studied. Seated at right, Claude Perrault points to a page of one of his animal monographs. To the left of the picture the secretary keeps the minutes. (From *Jardin des plantes* 1669, by LeClerc.)

else" through distillation, sublimation, and precipitation. Alchemy was becoming chemistry.

Helmont

The most important champion of "life chemistry" was Johannes Baptista van Helmont (1577–1644), a Flemish follower of Paracelsus, and an advocate of weapon salve. As a practicing alchemist, he believed he had succeeded in converting mercury into gold, and proudly christened his son, in commemoration of this great feat, Mercurius (Rosenfeld, 1985). But his enduring fame rests in his theory of animal fermentation as the fundamental process behind life chemistry. He believed that the many changes that take place in the body are guided by a special spiritual force, which he identified with the archeai of Paracelcus.

By far the most important of these changes is fermention, which produces an air that is identical to that given off by burning charcoal, and which can be distinguished from steam. Imaginatively,

Helmont discriminated between different kinds of "airs," which he, inventing a new term, named gases (from the Greek *chaos*): Gas sylvestre, or "wood gas," is the noninflammable gas released in belching, or when acid is pored on chalk; and flammable "gas pinque" is produced by putrefaction, and is released per anum (a diverting experiment). He became enthusiastic. Gas is universal, it is bound up with matter, spirit and soul, and is present in all substances. Indeed, it is a divine truth that can be visualized.

Although gas was the primary spiritual-like effector, Helmont considered water the primary substance. He arrived at this conclusion from his tree experiment. A five pound willow tree was planted in 200 pounds of dried earth. Five years later, having added only water, he found that the tree had increased in weight to 169 pounds, and that, the earth, reweighed after drying, had lost no weight. He concluded that the wood, bark, and roots had been formed from the water alone.

Helmont's most important contribution to physiology was his firm rejection of the Galenism that the chief function of respiration was to cool the extreme heat of the heart. He declared, to the contrary, that respiration supported animal heat. In his view, respiration produced the combination of an element of the venous blood with a "ferment" from air.

Again contrary to Galen, Helmont argued that it is an error to attribute heat as the effective agent in digestion, observing as evidence that cold fish digest their food. Digestion is not simply a matter of cooking, but is the chemical transmutation (alchemic overtones) of food into the parts of the body, and the true effector of this transformation is fermentation, which acts as an agent of "divine light," i.e., of God. He developed this theme with zeal, elaborating in detail on the six fermentations involved in converting food into tissue. There were three in the gut, where food is converted first into chlye, then into crude blood; one in the heart, where blood is made lighter; one in the arteries, where vital spirit is made; and the sixth fermentation takes place in the "kitchens" of the individual organs, where the final transformation takes place. His perception was that the body was an aggregate of tissue or organ metabolic kitchens, each with its own archeus (Rosenfeld, 1985). Noting that extracted blood froths like new wine, he thought that the vital fermentation that takes place in the left ventricle of the heart produces heat, which is distributed to the rest of the body (Mendelsohn, 1964).

Helmont liked the gut enormously: It is the engine of the primary biological process of fermentation (as evidenced in belching and farting), and, following this line of reasoning, he had the sensory-motive soul residing in the gut, the special honor being given to the pit of the stomach. His supporting arguments include noting that the heart does not stop immediately when a man has had his head blown off by a cannonball, but that a blow to the stomach can cause unconsciousness and cardiac arrest.

It has been suggested that Helmont's conviction in a stomach-residing soul was encouraged by his self-confessed experiences following swallowing the hallucinogenic herb Wolfsbane which contains the highly poisonous compound aconite (Hall, 1969a). Aconite increases membrane ion permeability, causes cardiac arrhythmia and respiratory depression, and could have given the sensation that his mid-region, rather than his head, was the center of understanding and imagination.

Sylvius

Franciscus Sylvius (Franz de le Boe or Francois du Bois) (1614–1672) further developed and refined Helmont's iatrochemistry. A German and a staunch Protestant, he held the chair of medicine at Leiden University. Although strongly influenced by Helmont, he was impatient with the idea of the mystical archeus as the force behind fermentation. Like Helmont, he regarded life as a chemical process, but he looked for straightforward explanations that were much more in line with current chemical concepts. He thought, for example, that the key to the identity between the chemistry of the living and the nonliving lay in the composition of salts and the properties of acids and alkalis.

Sylvius was particularly struck by the observation that mixing acids and alkalis often produced heat, and he thought that the corrosive properties of acids and alkalis were caused by some sort of fire contained in them. Animal heat came from an active acid-alkali fermentation in the right side of the heart, brought about by the mixing of chyle and lymph-filled blood with bile-laden blood (Underwood, 1972). However, as of old, he believed that respiration serves to temper this heat, to preserve its "ebullition."

Sylvius agreed with Helmont that digestion was fermentation, but he argued that physiological fermentation was actually the same process as the effervescence produced by vitriol (acid) on chalk. Saliva played an important role in this fermentation, acting not only in the mouth, but also in the stomach, as it accompanied masticated food (Foster, 1901). The next stage of digestion, which occurred in the small intestine, resulted from the effervescing/fermenting action of the combined

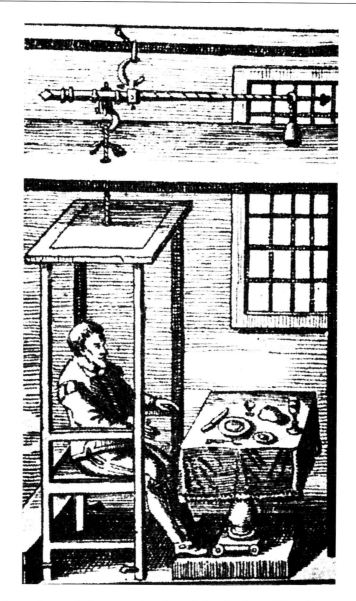

Fig. 12.2. Sanctorius's balance for carefully comparing the weight of solid and liquid ingested with that excreted, and changes in body weight. (From his *De statica medicina*, Venice, 1614.)

effects of the alkaline bile and acidic pancreatic juice. Not unexpectedly, he came to the conclusion that most diseases were the result of an excess of acid, and certain fevers were caused by too much alkali, and he constructed an elaborate acid-alkali therapy based on this principle.

Willis

The English anatomist and physician Thomas Willis (1621–1675) developed an iatrochemical theory of muscular contraction. In his work, *De*

motu musculari (1670), he proposed that animal spirits flow from the brain, along the solid nerve fibers, to the muscles. Arriving in the muscles, they came in contact with sulphurous and nitrous particles, which came from the blood, and a local miniexplosion ensued, dilating and shortening the muscle fibers.

IATROPHYSICS

The iatrophysicists scoffed at this old-fashioned approach. To them, all was physics and modern

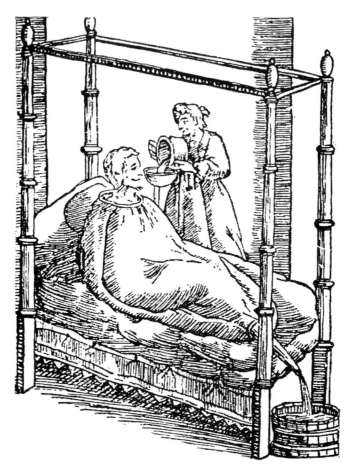

Fig. 12.3. Sanctorius's Balneatorium: a leather bag that allows flowthrough bathing of the patient in bed, and the collection of such fluid.

methods of measuring and calculating were of the essence. They rejected chemical explanations as fanciful, and, influenced by practical mechanics, applied the much more impressive mathematically based explanations of Newtonian physics to biology. Hypothetical machines were constructed, similar to popular mechanical automata, to account for animal function. The dominant themes were balance, harmony, and design; the favorite analogies were the clock, water-powered devices, and the steam engine. These models were wholly deterministic, and, once set in motion, resulted in an inevitable succession of events, which were also, in principle, reversible. The key concept was the balance of nature, which biological processes eternally revolve, like the seasons, but never progress.

Santorius

One of the most committed advocates of the quantitative process was Santorio Santorio (Sanctorius) (1561–1636), a professor at Padua University, who, over 30 years, routinely spent long periods eating and sleeping in a specially designed chair balance (Fig. 12.2), to obtain the difference in weight between what he ingested, solid and liquid, and what he excreted, solid and liquid (Eknoyan, 1999). He called this weight loss "insensible perspiration," and determined that the amount lost in one day, either through the pores in the skin or from respiration through the mouth, equaled one-half pound. He noted that this insensible perspiration can be seen by breathing on a glass. He also designed an enveloping, flowthrough bath (Fig. 12.3).

Sanctorius must have gathered an enormous amount of data, because he claimed, in a letter to Galileo, to have studied many thousands of subjects under the influence of different factors, such as air, food, age, movement, repose, sexual activity, and drink. However, all the records of his life of detailed recordkeeping were lost, and just a summary remains, in the form of aphorisms in his book, *De statica medicina (Concerning Static Medicine)*, published in 1614. But the book provides some idea of his meticulous approach: "In the first two hours after eating a great many perspire a pound or near; and after that to the ninth two pound ... sixteen ounces of urine is generally evacuated in the space of one night; four ounces by stool, and forty ounces upwards by perspiration" (Eknoyan, 1999).

Descartes

The French philosopher, mathematician, and physiologist, René Descartes (1596–1650), was the champion advocate of mechanical biology. Like many of his professional kind, even those of today, he was addicted to abstract, and somewhat dogmatic, speculation concerning biology. His basic tenant was that the composite can be fully explained from its simple components, and just as the universe is a machine acting in accord with physical laws (which we now fully understand), so is the body. He had mechanical explanations for motion and perception, and proposed a hydraulic model for motion, similar in principle to the piped water works that were popular at the time.

THE BODY-MACHINE

In *Traité de l'homme (Treatise on Man)* (1662) Descartes produced a rational explanation of body function consistent with most modern ideas in mechanics, but he carefully maintained a clear distinction between life processes and the governing soul: "All bodily functions must therefore act according to mechanical laws and the human body is simply a machine ruled by a rational soul" (Hall, 1972). The difference between a living and a dead body is similar to that between a working, wound-

Fig. 12.4. Descartes' hydraulic model of the brain causing muscle contraction. The pineal (**H**) secretes animal spirit into the ventricle (**E–E**), which then makes its way through a network of small pores (a) through the walls of the ventricle (**A**). The spirit passes down tubes (**B**) to the spinal cord (**D**), and hence to the muscle, which swells causing contraction. Changes in the inclination of the pineal controls the amount of animal spirit flowing to the muscle.

up watch and a stopped or broken watch. In his monumental, *Meditationes de prima philosophia...* (1641), he declared that the human body is "a machine so built and put together of bone, nerve, muscle, vein, blood and skin, it would not fail to move in all the same ways as present." He elaborated on his human-machine model with great enthusiasm:

"And as a clock composed of wheels and weights observes not less exactly all the laws of nature when it is ill-made and does not tell the hours as well as when it is entirely to the wish of the workman, so in like manner I regard the human body as a machine so built and put together of bone, nerve, muscle, vein, blood and skin, that still, although it had no

Fig. 12.5. Descartes' model of sensory processing. Light from the arrow enters the eyes and forms an inverted image on the retina. The image travels along the hollow nerves to the pineal gland, where it is interpreted (*Treatise on Man,* 1662).

mind, it would not fail to move in all the same ways as at present, since it does not move by the direction of its will, nor consequently by means of the mind, but only by the arrangement of its organs."

MECHANICAL MODELS

Descartes dogmatically maintained that the mechanical view is the only one necessary:

"I say, that these functions imitate those of a real man as perfectly as possible and that they follow naturally in this machine entirely from the disposition of the organs—no more nor less than the movements of a clock or other automaton, from the arrangement of its counterweights and wheels. Wherefore it is not necessary, on their account, to conceive of any vegetative or sensitive soul or any other principle of movement and life than its blood and its spirits, by the heat of the fire which burns continually in its heart and which is of no other nature than all those fires that in inanimate bodies."

Descartes agreed with Harvey's description of the circulation of the blood, but strongly opposed Harvey's explanation that blood, was pumped by the contracting heart. This, he complained required that the heart walls have the mysterious innate property of contraction and expansion. Descartes was particularly proud of his more rational "steam engine" explanation of the heart's motion, which depended on intense heat in the heart, the "heat without fire." In *La description du corps humain* (1664), he writes that this heat "is like a mainspring and principle of all the movements that are in it." Blood, coming from the liver, falls drop by drop into the hot right cavity, where it is vaporized, dilates the heart, and expands to the lungs, and hence to the whole body. In the lungs, some blood "vapor" is condensed and falls, drop by drop, into the left cavity of the heart to nourish the fire that exists there. This movement of blood in and out of the heart is coordinated by the valves (Harvey's "little doors"), which act as one-way gates. Descates complains dismissively of "those who do not know the force of mathematical demonstration and are

A

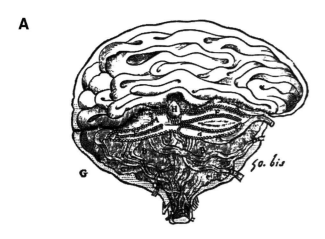

B

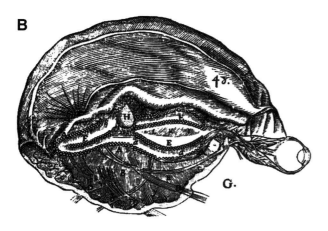

Fig. 12.6. Descartes' mechanical-hydraulic model of sleep. (**A**) represents sleep, (**B**) awakefulness. In sleep, the flow of animal spirit from the pineal (**H**) has been reduced to a trickle, the ventricles have collapsed, and the hollow nerves are flaccid. When awake, copious amounts of animal fluid are produced, swelling the brain and making the nerves taut (*Treatise on Man*, 1662).

unaccustomed to distinguish true reasons from merely probable reasons."

BRAIN AS RESERVOIR

Descartes saw the brain as a reservoir for the fluid humors, and the nerves as distributing pipes. In his system, the heart is the source, or fountain, of animal spirits, and the arteries carry the "most agitated and most lively particles" from the heart to the brain ventricles, "something like air or a very subtle wind." As illustrated in Fig. 12.4, Descarte's

brain was geometrical, and there was no attempt to be anatomically accurate: The pineal (H) secretes spirit into the ventricle (E–E); this spirit makes its way through a network of tiny pores (a) in the walls of the ventricle (A), along threads (B), and down the spinal cord (D). Changes in the inclination of the pineal cause changes in the flow of spirit down the spinal cord.

HYDRAULIC MODEL FOR MOVEMENT

The ventricle fluid is conveyed to the muscles down the hollow nerve pipes, which act like water pipes in hydraulic machines, causing the muscle to expand: hence, contraction. "The sanguine spirits enter the cavities of the brain" (the ventricles), and from there,

> "...enter the pores (or conduits) in its substance, and from these conduits proceed to the nerves. And depending on their entering (or their mere tendency to enter) some nerves rather than others, they are able to change the shapes of the muscles into which these nerves are inserted and in this way to move all the members. Similarly you may have observed in the grottoes and fountains in the gardens of our kings that the force that makes the water leap from its source able of itself to move divers machines and even make them play certain instruments or pronounce certain words according to the various arrangements of tubes through which the water is conducted."

MECHANICAL MODEL OF MENTAL FUNCTION

The sanguine spirits "dilate the brain rendering it fit to receive impressions from the soul to be the organ of common sense, of imagination and of memory." This process of sensory perception is illustrated in Fig. 12.5. Inverted images of the arrow are projected onto the retina of each eye. The images are carried down the hollow optical nerves, by animal spirit, to the pineal gland, where they are united into a single upright image.

Descarte's hydraulic model of sleep is shown in Fig. 12.6. During sleep, the pineal (H) only pro-

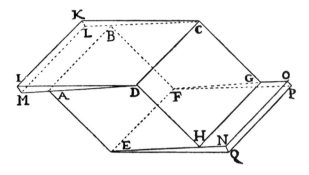

Fig. 12.7. Structure of muscle according to Steno's geometric model. A single muscle is viewed as a parallelogram (ABCDEFGH). The tendons at either end are prisms (DMIKLC and EFQPON). (From Bastholm, 1950.)

duces a trickle of animal spirit; the ventricles are consequently collapsed, the brain flaccid, and the nerve filaments (D) hang limply. But, in the awakened state, the pineal produces an abundance of animal spirit, the brain becomes turgid, nerves are taut and the muscles are tense.

THE HOUSE OF THE SOUL

According to Descartes, the rational soul, which was the entity that ran the mechanical body, was housed in the pineal gland, an appropriately small, single body at the base of the brain, behind and between the eyes (Figs. 12.4 and 12.5). Here it acts "like the turncock who must be in the main to which all the tubes of these machines repair when he wishes to excite, prevent, or in some manner alter their movements." This separation of the mind and body was the basis of Descarte's extremely influential philosophy of dualism.

Descartes had used some modern discoveries in constructing his body-machine, and invented others. He was not constrained by actual anatomical detail, but was more interested in the completeness or beauty of his model, its "clear and perspicuous ideas." His model explained movement, blood circulation, sleep, and sensory perception; he was impatient with the lack of evidence or anatomical knowledge to support his beautiful scheme. Perhaps he thought that later discoveries would provide the missing facts.

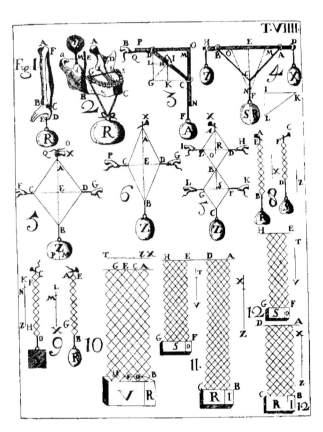

Fig. 12.8. Borelli's geometric models of muscle mechanics. The contracting elements, the "finest fibers" of muscle are viewed as rhombs. (From *De motu animalium*, 1685.)

Steno

The Danish physician Niels Stensen (1638–1686), or Steno, proposed a strictly mechanical and geometric model of muscular contraction. The muscle was considered to consist of long fibers with transversely connecting filaments, arranged in parallelogrammatic fashion. Contraction was the result of a change in the angles of the parallelograms (Fig. 12.7), though how this was brought about was not clear.

Borelli

Giovanni Alfonso Borelli (1608–1679), a follower and pupil of Galileo, was an accomplished mathematician, anatomist, and physiologist. He was an outstanding exponent of the mechanical approach, and his models are remarkable in their boldness and imagination. His approach has a modern flavor, in that many of his ideas invited testing.

In his most famous book *De motu animalium* (1679), Borelli expounded a physical-mathematical science of movement in animals. The first part is a geometrical treatment of movements of individual muscles, and groups of muscles in humans and animals, and deals with the mechanics of posture, flight, and swimming.

The second part concerns the cause of muscular contraction. Borelli disagreed with the generally accepted view that the tendons were the contracting elements of muscle, the flesh being merely a passive filler. He argued that, to determine how muscles worked, we must examine their structure, dissect out their parts, and closely observe their action.

Believing that the contraction mechanism of muscle would be common to all species, he undertook a pioneering, detailed, comparative anatomy of human, fish, and even insect muscles. The volume changes in muscle contraction were produced through its fine structure of rhomboid-shaped fibers (Fig. 12.8). He tested the theory that the increase in muscle volume during contraction (as was thought at the time) was caused by a balloon-like swelling from spiritous fluid pouring out of the nerves, or from a gaseous explosion, as proposed by Willis. He submerged a live animal in water and slit its muscles, arguing that, if some "spiritous gas" had entered the muscle, it should "burst forth from the wound and ascend through the water" (Eldon, 1965). No bubbles came from the contracting muscles. His conclusion was that "inflation results from something in the muscles themselves, that such an action is possible is rendered clear by innumerable experiments which are continuously being made in chemical elaborations as when spirits of vitriol are poured on oil of tartar; ... at once boil up with a sudden fermentation."

Borelli's alternative hypothesis was that a sudden fermentation was the effective cause of contraction, and that the explosion, which caused the muscle to increase in bulk, was triggered by the release of a drop of nervous fluid into the muscle,

the "succus nerveus," which had a "liquid constituancy like spirits of wine" (Foster, 1901).

More imaginatively, Borelli suggested a mechanical model that allowed the nerve to transmit both from and to the brain. He envisioned the nerve fibers as canals filled with spongy pith-like material, and turgid with succus nerveus. When this swollen tube is struck or pinched, "spiritious droplets" are squeezed out into the appropriate muscle, "whence ebullation and explosion follows, by which the muscles are contracted and rendered tense" (Eldon, 1965). On the other hand, when the sensory nerves are compressed or struck, an "undulation" is conveyed to the brain, and hence to the faculty of the sensitive soul. He was also interested in the mechanics of blood, and appreciated that a continuous flow of blood from the arteries to the veins depended on the capillaries having elastic walls.

Borelli was intrigued by the watch analogy of the body-machine hypothesis, and, taking the analogy a step further, he argued that the body had need of a regulatory mechanism. "Whence as in a watch, so in the animal-the natural automaton-a regulatory device must be added which by mechanical necessity will bridle the motive force lest it transgress the laws set by the Divine Architect" (Foster, 1901). He speculated that this regulator is in the air, and that, in some way, air particles, which acted like "little aerial machines," mixed in the blood during respiration, where they "perform an oscillatory movement in the manner of the pendulum." This is a reason why air is necessary for life.

Borelli applied mechanical principles to other physiological problems. For example, he regarded digestion as primarily a mechanical process, observing that, in certain birds, "the crushing, erosion and trituration of food is effected by the muscular stomach (the gizzard) itself" (Foster, 1901). He introduced "glass globules, or empty vesicles and leaden cubes pyramids of wood and many other things," into the stomachs of turkeys, and found, the next day, that "the leaden masses crushed and eroded the glass pulverized and the remaining ingesta in the same condition" concluding that the action of the teeth and the stomach are the same.

Borelli dismissed the idea that body heat was produced by the heart. Instead, he proposed, like many others, that body heat resulted from the frictional resistance produced by blood flow through the tissues. It followed that, the greater the blood velocity, the greater the amount of heat produced, and, since blood flow is swiftest in the heart, the heart should be the warmest organ. But, using one of the new thermometers, he found that the temperature in the left ventricle of a stag was no warmer than its liver, lungs, or intestines.

Other Iatromechanists

The concept that animal heat is the result of friction was held until the 1800s, although there was some disagreement over the site where it was generated. For example, Archibald Pitcairn (1652–1713) held the extreme iatromechanist view that the innate heat produced by blood, meeting resistance "from the sides of the arteries,…may be considered as a rectangle under the Velocity and the Distance" (Mendelsohn, 1964). Meanwhile, Haller believed that body heat was derived from the friction of blood in the heart, and Hermann Boerhaave (1668–1728) thought that the frictional heat was made in the lungs. But did not the lungs also serve to cool the blood? The reason for this apparent contradiction, Boerhaave declared, was that, if unabated, the heat produced in the lungs would endanger life.

Giorgio Baglivi (1668–1707) summarized the ultimate iatromechanist position in his *Opera ominas medico-practica et anatomica,* published in 1704:

"Whoever examines the bodily organisms with attention will certainly not fail to discern pincers in the jaw and teeth: a container in the stomach; water mains in the veins, the arteries and the other ducts; a piston in the heart; sieves or filters in the bowels; in the lungs , bellows; in the muscles the force of the lever; in the corner of the eye a pulley, and so on" (Nuland, 2000).

He scathingly dismisses the obscurantism of the iatrochemists:

"So let the chemists continue to explain natural phenomena in complex terms such as fusion, sublimation, precipitation etc. … The natural functions of the living body can be explained in no other way so clearly and easily as by means of the experimental and mathematical principles with which nature herself speaks" (Translation from Nuland, 2000). However, the iatrochemists and iatrophysists share one important common approach that was to prove an engine for the further development of physiology—they attempt to explain life processes solely in terms of chemistry or physics.

13 New Technology, New Physiology

The development of new instruments facilitated radical breakthroughs in scientific concepts. The improved thermometers of the seventeenth century allowed scientists, for the first time, to properly distinguish between temperature and heat; the new gas pumps showed that air changed in volume during respiration; and the microscope revealed a whole new part of creation, never seen or imagined before.

THE NEW WORLD OF THE VERY SMALL: THE MICROSCOPE

The new technology of "ocular magnification" caused a virtual revolution in our perspective of the living world. The most impressive explorers of these new territories were Hooke, Malphigi, Leewenhoek, and Swammerdam.

Hooke

The fresh world revealed by the microscope caused something of a sensation. The year 1665 saw the publication of Robert Hooke's (1635–1703) richly illustrated book, *Micrographica* or *Some Physiological Descriptions of Minute Bodies, Made by Magnifying Glasses, with Observations and Enquiries Thereupon.* It was eagerly read. For example, Samuel Pepys commented in his diary, "Before I went to bed I sat up till two o'clock in my chamber reading Mr Hooke's *Microscopical Observations,* the most ingenious book I ever read in my life."

Micrographica was essentially a picture book, with representations of magnified bits of animals and inorganic materials, such as the edge of a sharp razor and ice crystals. Hooke had a particular interest in insects, especially fleas and lice (both readily available), flies, moths, and bees. At the time, the book was particularly famous for its illustration of a monstrous flea (Fig. 13.1), but Hooke is primarily remembered for his description of "little boxes or cells" in cork. Hooke realized that the microscope was a technological breakthrough (Fig. 13.2) and reasoned that it would correctly and simply establish the truth of the mechanical philosophy, and would take us down to the realm of philosophical essences, allowing us to determine "all the secret workings of nature."

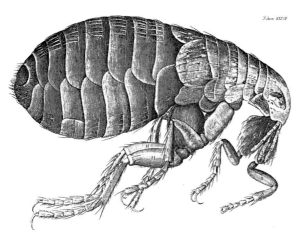

Fig. 13.1. Hooke's famous flea. (From *Micrographica*, 1651.) Samuel Pepys commented in his diary, "Before I went to bed I sat up till two o'clock in my chamber reading Mr Hooke's *Microscopical Observations,* the most ingenious book I ever read in my life."

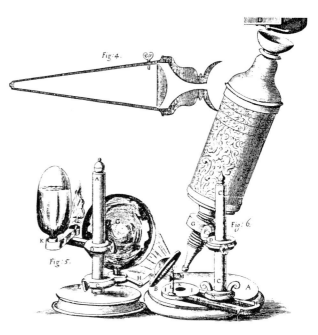

Fig. 13.2. Hooke's microscope. (From *Micrographica*, 1651.) A technological breakthrough. (National Library of Medicine.)

Malphigi

Marcello Malphigi (1628–1694), a friend of Borelli, and personal physician to Pope Innocent XII, published many short reports in the *Proceedings of the Royal Society of London*, which displayed astonishing new findings on a wide variety of subjects. However, his first and greatest experiments were published in a book, *De pulmonibus* (1661), in which he described the small anatomy of the lung. It was thought at that time that the lung was composed of undifferentiated parenchyma in which air and blood freely mixed. Malphigi showed that the lung was not merely a "fleshy" tissue, but that it had a distinctive microstructure, consisting of tiny inflatable membranous sacs (alveoli). His most important physiological discovery followed quickly. Looking at the magnified lung of the living frog, he saw blood streaming through smaller and smaller branches of blood vessels, which he called "tubules," not visible to the naked eye. Then he observed that the blood streams reunited in larger and larger veins. To quote his own description,

"I saw the blood, showered down in tiny streams through the arteries, after the fashion of a flood, and I might have believed that the blood itself might have escaped into an empty

space and was gathered up again by a gaping vehicle, but an objection to the view was afforded by the movement of the blood, being tortuous and scattered in different directions and by its being united again in a determinate part" (Castiligioni, 1958). His "doubt was changed to certainty" on examining dried frog lungs under his microscope. He confirmed that this blood was only to be found in vessels, and did not seep out into the surrounding flesh, as had been previously believed. "Hence it was clear to the senses that ... the blood was not pored into the spaces, but was contained within tubules" (Foster, 1901). He had discovered the capillaries that connect the arterial to the venous systems, and so provided the missing link in Harvey's circulation theory (Fig. 13.3).

In his book *De cerebro* (1665), Malphigi showed that he could separate the white matter of the nervous system into bundles of fibers, and traced their connection to the brain via the spinal cord. In a comparative study of the body organs, in *De*

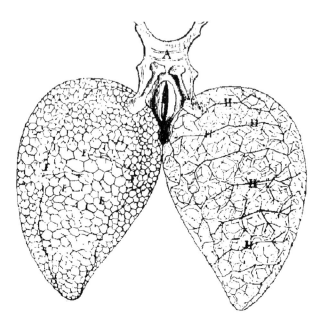

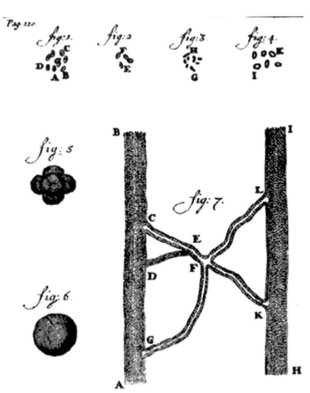

Fig. 13.3. Frog lungs showing capillaries (**HHH**). (From Malphigi, *De pulmonibus*, 1661.) Malphigi had discovered the connection between the arterial and venous blood systems, and so provided the missing link in Harvey's circulation theory.

Fig. 13.5. Leeuwenhoek's illustrations of red blood corpuscles. (**1**) Salmon; (**2–4**) eel; (**5–6**) aggregation of eel blood corpuscles. Fig. 7 (*inset*)shows the capillaries (**EF**) between an artery (**HI**) and a vein (**AB**). (National Library of Medicine.)

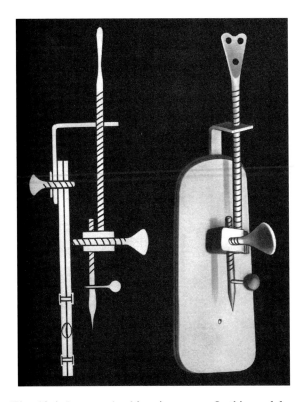

Fig. 13.4. Leeuwenhoek's microscope. In this model, a biconvex lens is mounted between the metal plates. The object to be viewed is secured on the metal point.

viscerum structura (1666), he showed that the liver had the same secretory microstructure in all of the vertebrates he examined, from fish to humans, and he demonstrated that a duct connected the gallbladder to the liver, which suggested to him that bile was not secreted by the gallbladder, as had been believed by the ancients, but is stored there. This he confirmed by experiment by tying off the duct in a live kitten, which caused the gallbladder to remain empty of bile. In a detailed examination of the kidney structure, he found that the kidney was composed of tubules and tiny blood vessels, and that, in the kidney cortex, the tubules had blind ends that were surrounded by a sphere of fine blood vessels, the malphigian bodies. His careful dissection of insects revealed many new features, including spiracles and air tubes, and the multichambered insect heart and their specialized excretory organs, the malphigian tubules.

Malphigi's most sensational discovery, which was communicated to the Royal Society in 1675, was the existence of a whole new world of tiny living creatures, invisible to the naked eye, which he had found in a tub of water that had stood for a few days. He estimated that these "animalcules" were more than 10,000 times smaller than the water flea. Members of the Royal Society attempted to reproduce this incredible claim, but their first attempts, using "common pump water," failed. Finally, using a $\times 100$ magnifying glass on pepper-infused water that had remained undisturbed for nine or ten days, the tiny creatures were detected. The hidden, unimagined world of teeming microscopic life had been discovered and verified.

Leeuwenhoek

Antoni van Leeuwenhoek (1632–1723) received no formal training in science, and knew no Latin, so could not read the classical texts. His microscopes consisted of a double-convex lens held in a brass frame, with a movable silver needle, to which the specimen was attached (Fig. 13.4). He ground his own lenses, and one single lens reputedly had a magnifying strength of around 300 times. Like some later scientists, he was very secretive about his "technology," wishing to keep the advantage to himself. Leeuwenhoek communicated some 300 letters of astonishing detail, on a wide variety of subjects, to the Royal Society, over a period of 50 years. He was the first to note the striated nature of muscle and the rods and cones in the retina, and gave the first accurate descriptions of the red corpuscles found in blood of fish, frogs, birds, and humans (Fig. 13.5). He saw "many little animalcules" in pond water, and in 1702 he gave the first description of the behavior of a ciliate (*Vorticella*):

> "In structure these little animals were fashioned like a bell, and at the round opening they made such a stir, that the particles in the water thereabout were set in motion thereby.... And though I must have seen quite

20 of these little animals on their long tails alongside one another very gently moving, with outstretched bodies and straightened-out tails; yet in an instant, as it were, they pulled their bodies and their tails together, and no sooner had they contracted their bodies and tails, than they began to stick their tails out again very leisurely, and stayed thus some time continuing their gentle motion: which sight I found mightily diverting."

Leeuwenhoek made microscopic observations on the plaque between his own teeth, and reported that he saw "with great wonder, that in the said matter there were many very little living animalcules, very prettily a-moving." One sort which was found in great numbers "oft-times spun round like a top," and the biggest sort "bent their body into curves in going forwards." He had discovered bacteria.

Leeuwenheok also examined semen under his microscope, got not by "sinful contrivance," but obtained by hastening from the marriage bed to the microscope (Wilson, 1995), "before six beats of the pulse had intervened." To his astonishment, he saw a great number of living eel-like creatures, a million of them would not equal in size a large grain of sand. What was the role of these semen or "sperm animals," these spermatozoa? Were they signs of illness or were they products of corrupted or putrefying semen (like the tiny animals produced in old water)?

His thinking continued to advance. For example, he took a position on spontaneous generation, based on his observations on the reproduction of rotifers and insects and the spawning of mussels. And he wrote, in 1702, "Can there even now be people who still hold to the ancient belief that living creatures are generated out of corruption?" Leeuwenhoek's discoveries brought him considerable celebrity, and he was invited to explain and demonstrate his microscope by George I of Britain and Peter I of Russia. At his death, he left 247 finished microscopes and 172 lenses.

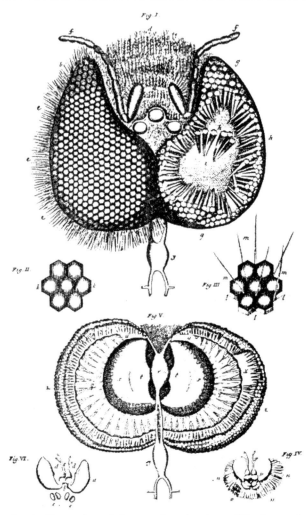

Fig. 13.6. Swammerdam's illustration of the amazing structure of the honey bee's compound eye.

Swammerdam

Jan Swammerdam, (1637–1680) was a deeply religious morphologist, so much so that his research has been termed "physiotheology" (Bäumer, 1987). He believed that he could best serve God by studying the secrets of nature. The power of God was demonstrated in the smallest of animals, in that the tiny insects were no less perfect than higher animals, a feature that could only have been brought about by God himself. This position was supported by the philosopher/mathematician Gottfried Wilhelm Leibnitz (1646–1716), who declared that what the telescope takes away, the microscope returns: "Nothing better corroborates the incompa-

rable wisdom of God than the structure of the works of nature, particularly the structure which appears when we study them closely with the microscope" (*Reflections on the Common Concept of Justice*).

Swammerdam developed a technique of injection to reveal microscopic structure, and made especially brilliant dissections of insects and mollusks. He published a *General History of Insects* in 1667, and a detailed account of the morphology of the mayfly in 1675. He devoted much of the last five years of his life to dissections of the bee and other insects, and other small animals, including the tadpole and the snail. Over 50 years after his death, these manuscripts were published by Boerhaave in a large and appropriately named, *Biblia naturae* (*The Bible of Nature*, 1737). In this book, Swammerdam gave the first accurate descriptions of the insect compound eye (Fig. 13.6) and the bee stinger. He was also the first to discover flat, oval, nucleated red blood cells in frog blood.

Swammerdam was also an experimentalist, and performed an elegant series of experiments on contracting muscle. He was able to demonstrate that muscle did not increase in volume on contraction, by placing isolated muscle into a water-filled vessel, and noting that, when the muscle contracted, there was no change in the level of the water.

VITAL AIR: THE DEVELOPMENT OF PUMPS AND EARLY THEORIES OF RESPIRATION

During the seventeenth century, new and improved air pumps awakened an interest to discover the properties in air that are essential for sustaining life.

Boyle

Robert Boyle (1627–1691) made important observations with his improved vacuum pump. He studied the effects of decompression on a variety of animals, including kittens, larks, linnets, green finches, frogs, and toads, and was able to demonstrate that, in an evacuated chamber, flame and life were extinguished more quickly, indicating that air was necessary for both. He did not believe that this air was used to cool the overheated heart or blood, observing that cold frogs need to respire. "There is some use of the air, which we do not yet so well understand, that makes it continually needful to the life of animals". But he refused to speculate: "What the air does in respiration … I choose to confess a careful ignorance than to profess false knowledge." He believed that air was necessary for the respiration of aquatic animals. "Air lurks in water … and of it fishes may make some use by separating it when they strain the water through the gills."

Characteristic of his time, Boyle was also deeply religious. In his essay, *The Christian Virtuoso*, he

argued that science can be perverted to discredit religion, but an increase in knowledge would rule out that the world is produced by "blind chance." In his will, he bequeathed an annual lecture series to prove the Christian religion, "against notorious infidels, viz Atheists, Theists, Pagans, Jews and Mohametans" (Lewis, 1994).

Hooke

Robert Hooke, whom we already discussed as the famous microscopist, was also an assistant to Boyle at one time, and built some of his vacuum pumps. Hooke took Boyle's biological observation further. His most famous series of experiments concerned investigations into the function of the act of breathing. He artificially respirated a dog with a bellows inserted into the lung, through-ventilation being afforded through small holes that had been made in the lung. He noted that "Upon ceasing this blast and allowing the lungs to lye still, the dog would immediately fall into dying convulsive fits; but he as soon revived again by renewing the fullness of his lungs with the constant blast of fresh air." Hooke concluded, in his contribution to the Royal Society (*Philos Trans Royal Soc 1*, 1665–1666 p. 536–540): "Which seems to be Arguments that as the bare motion of the Lungs without fresh air contributes nothing to the Life of the Animal, he being found to survive as well, when they were not moved, as when they were; so it was not the subsiding or movelessness of the Lungs, that was the immediate cause of death, or the stopping the Circulation of the Blood through the Lungs, but the want of a sufficient supply of Fresh Air."

Hooke speculated that the use of respiration was to ventilate the blood. During its passage through the lungs, the blood is "disburdened of those excrementitious streams proceeding from the most part from the superfluous serosities of the blood" (*Tractatus de corde*, 1669) (Cournand, 1982).

Hooke performed an apparently successful transfusion of blood between dogs which was mentioned in the *Diary of Samuel Pepys* (14 November 1666): "This did give occasion to many pretty wishes as of the blood of a Quaker be let into an archbishop, and such like."

Lower

Richard Lower (1631–1691) a pupil of Boyle, investigated the function of the lung. He found that blood in the pulmonary vein was already red, before it had reached the heart, and, using Hooke's artificial respiration technique, he further discovered that this reddening of the blood "must be attributed entirely to the lungs, as I have found that the blood which enters the lungs completely venous and dark in color, returned from them quite arterial and bright." He suggested that "this red color is entirely due to the penetration of particles of air into the blood" (*Tractatus de corde*, 1669) (Cournand, 1982).

Mayow

The imaginative and original John Mayow (ca. 1643–1679) took the next step in determining that it was not air per se, but some part of air that was necessary for the burning of flame and sustaining life. He placed an animal or burning object under an inverted cupping glass with its rim immersed in water, and noted that the water in the glass rose. This, he thought, was because the air was becoming less springy or elastic, because it was giving up a spirit portion that contained flame "food," the igneo-aerial spirit. Potassium nitrate was long known for its explosive powers, and, since lightning and thunder also possessed this property, it was thought that some sort of nitre exists in air.

Mayow believed that his atmospheric nitro-aerial particles are indispensable for the production of fire, and that "animals and fire draw particles of the same kind from air." He wrote "Respiration consists furthermore in the separation of the air by the lungs, and the intermixture with the blood mass of certain particles absolutely necessary to animal life, and the loss by the inspired air of some of its elasticity (in contemporary terms, its oxygen). The particles of the air absorbed during respiration are designed to convert the black or venous blood into

the red or arterial" (Mayow, via *Oxford Bibliographical Society,* 1908. Mayow contended that water contains nitro-arterial spirit, and that the gills of fishes function as lungs to extract spirit particles from water. When water is boiled, the vital substance is driven off, and the fish dies. Developing the same line of enquiry, he thought that the placenta of the human fetus functioned as a "uterine lung" to obtain nitro-arterial particles from the mother. In experiments with blood, he noted that in the vacuum pump, venous blood effervesces mildly, but that arterial blood bubbles more freely. This ingenious scientist made many seminal discoveries concerning respiration, and may have been on the road to discover oxygen, but he died at the early age 36 years, and his work fell into oblivion. As discussed later, the erroneous phlogiston theory of combustion was almost universally adopted for the next 100 years.

THE THERMOMETER AND BODY HEAT

Since ancient times, warmth has been associated with life; cold, with death. The source or site of this vital heat were old questions, still debated in the 17th century. For Harvey, the "innate fire" was intimately linked to the blood—the very essence of blood was heat—and the liver, spleen and lungs were thought to be especially hot due to their large supplies of blood (Frank, 1980). Descartes thought that the intense heat that powered his mechanical model of circulation was produced in the heart. Helmont believed that heat-generating fermentation occurred in the left ventricle of the heart. Boerhaave theorized that heat was made in the lungs through the friction produced by transpulmonary blood flow.

How to decide? Fernel claimed that he could feel excessive heat when he placed his fingers into an opening in the chest cavity of a live animal. But when Lower inserted a finger into the heart of an animal during a vivisection, he detected no greater warmth. Borelli agreed with Lower, he had no sensation of burning heat when he placed a finger into a chest opening. Clearly the finger was an unsatisfactory tool for detecting different degrees of warmth.

In the early 1600's, heat-measuring devices that relied on the change in volume of a trapped bubble of air—thermoscopes—were elaborated. Galileo constructed the first one around 1600, and Santorio used a thermoscope to measure body temperature. But thermoscopes proved impossible to calibrate precisely because the trapped gas volume was susceptible to changes in barometric pressure. A device that measured the change in volume of liquid enclosed in a glass tube, the thermometer, was much more satisfactory. Many schemes were proposed for calibrating thermometers with fixed reference points. Sir Isaac Newton proposed a scale with freezing water as the lowest point and body temperature as the upper, the scale to be divided into 12 degrees. Daniel Gabriel Fahrenheit (1668–1736), a student of Boerhaave with a special interest in biological heat measurement, invented the mercury thermometer around 1714. In Fahren-heit's scale, zero was the coldest temperature that he imagined could be on earth, obtained through a mixture of ice and salt. The upper calibration point , marked 90°, was body temperature got by placing the ther-

mometer under tongue or arm pit. In this scale, water boiled at 212° and froze at 32°. When boiling water was later used as the revised upper calibration point, normal body temperature was adjusted to 98.6 F. Fahrenheit's scale was widely accepted, especially in the English-speaking world. Today, however, Anders Celcius's (1701–1744) simpler scale, which set ice at 0° and boiling water at 100° at one atmosphere pressure, is now almost universally adopted, with the USA a notable exception.

The thermometer was quickly used to settle old disputes. Borelli placed a thermometer in the left ventricle, liver, lungs and intestines of a living stag, and could find no difference in measured temperatures. Hales measured how much and how quickly blood heated inspired air by comparing the temperatures of air inhaled through the nose with air exhaled out the mouth (Mendelsohn, 1964).

These findings were interesting but not radical. The truly important breakthrough that the thermometer provided was that, for the first time, scientists could distinguish between the degree of heat (temperature), and the amount of heat (heat content). For physiology, this meant the birth of the discipline that would become known as bioenergetics (*see* Chapter 15).

14 Reaction and Opposition

The increasing encroachment of science into what had been undisputedly religious territory provoked a wave of resentment and opposition in the late seventeenth century.

Although the "new world discovered" represented by the explosive growth of scientific inquiry was news and wonder to the general educated public, it was of serious interest—or, more accurately, only understood by—a comparative few. For the great majority of the literate, who had been educated almost exclusively in the classics, a good knowledge of Greek and Latin language and literature alone distinguished the educated gentleman, and made him fit for office. For example, the wise and esteemed Dr. Samuel Johnson dismissed science in his chapter on Milton in the *Lives of the Poets*: "But the truth is that knowledge of external nature, and the science which that knowledge requires, are not the great and frequent business of the human mind." That great business is reserved for "those authors … that supply most axioms of prudence, most principles of moral truth, and most materials for conversation; and these purposes are best deserved by poets, orators and historians." However, although an outstanding figure in English literature, Dr. Johnson's mind was not overly encumbered with knowledge of "external nature." In Boswell's *Life of Johnson*, Johnson is quoted as saying, "Swallows certainly sleep all the winter. A number of them concobulate together, by flying round and round, and then all in a heap throw themselves under water, and lye in the bed of a river."

An antiscience reaction became fashionable in late seventeenth century England, and increasingly vicious attacks were mounted on the Royal Society and its members (Ashton, 1883; Stinton, 1968). The amateur scientists, called "virtuosi" (the term "scientist" was not coined until the nineteenth century), were directly condemned. For example, in 1670, a Dr. Stubs complained that the members of the Royal Society were "very great impostors or men of little reading (i.e., classics)." They were "attempting to overthrow the universities as idiots and ignoramus's…destroy the established religion and to involve the nation in popery." Stubs wrote vigorously to Robert Boyle "I believe that not one

Fig. 14.1. Hogarth's parody of anatomists in *"The Reward of Cruelty,"* 1734. Cupidity, ignorance, and stupidity are well illustrated.

lives that doth not condemn your experimental philosophy. The most common complaint against science is of the nature. "This fanatical interest in what a microscope reveals is preposterous! What does it matter what the eye of a fly looks like? What possible use is this knowledge?—It is a waste of a serious gentleman's time."

Scientists were scoffed at and ridiculed. In Samuel Butler's poem, "The elephant on the moon," written sometime in the 1670s, a group of virtuosi, observing the moon with a long telescope become excited about seeing battling armies and an elephant on the moon. They are shamed by the

"common sense" of a footman who shows the elephant to be a trapped mouse and the armys to be insects crawling inside the tube of the telescope.

Shadwel's play, *The Virtuoso* (1673), was a scathing attack in which the virtuoso, Sir Nicholas Gimcrack, is presented as "a crank, superstitious and gullible, and interested only in the eccentric or monstrous; a sham philosopher, vain and shallow, whose ostensible love of learning was at root but idle curiosity, and whose learning itself was studiously divorced from practical reality" (quoted in Stinton, 1968). The play makes fun of such idiocies as the transfusion of sheep's blood into a mad-

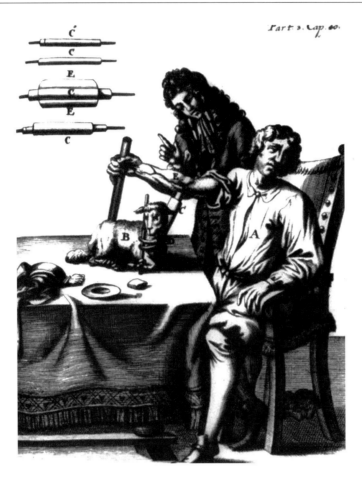

Fig. 14.2. Blood transfusion from a sheep to a man. (From Purmann's book on military surgery, 1721). (National Library of Medicine.)

man (such an experiment had actually been performed (Fig. 14.1), and the folly of spending 2000 pounds sterling on microscopes to study such trivia as the nature of mites in cheese.

The periodical, the *Tatler* (No. 216), makes up a will for Gimcrack, with such items as "To my little daughter Fanny, three crocodile eggs and ...the nest of a humming bird. My eldest son John having spoken disrespectfully of his little Sister, whom I keep by me in Spirits of wine ...I do disinherit." The famous essayist, Joseph Addison, joined in the attack: "When I married this Gentleman he had a very handsome estate, but upon buying a set of Microscopes he was chosen a Fellow of the Royal Society, from which time I do not remember him speak as other people do, or talk in

a manner that any of his family could understand" (*Tatler*, No. 221).

This wave of antiscience satires lasted into the next century. The third part of Jonathan Swift's *Gulliver's Travels* (1726) ridicules the Royal Society as the Grand Academy of Lagado, in which odd and eccentric members are depicted as obsessed with absurd projects. One of the professors had been "eight years upon a project for extracting sunbeams out of cucumbers," and reckoned he needed eight years more to succeed. Another's task was to discover how to "reduce human excrement to its original food" by "removing the tincture" and "scumming off the saliva," and so on. And a third was researching a new treatment for stubborn colic by pumping air into the gut with a bellows. When

Fig. 14.3. Caricature of the dissection room. (From Rowlandson, 1756–1827.) The main characters are actual contemporary scientists. The atmosphere is one of "unhealthy" obsession.

he had pumped up enough, he expected that the "adventitious wind would rush out, bringing the noxious along with it." To his consternation, the experimental dog "died on the spot."

The attacks are not just in literature, but in art, also. Hogarth's engraving of the dissection of the bad apprentice (Fig. 14.2) appears to be a parody of the frontispiece of Vesalius's *Fabrica*. Human dissection was an enduringly popular subject for sensational caricature (Fig. 14.3). The price of unconventional opinions could be heavy (Fig. 14.4).

Some conservative scientists were also becoming uncomfortable at the increasing power and scope of science to provide materialistic explanations for life. One of the most influential of these was the famous German chemist and physician,

Georg Ernst Stahl (1660–1734), who strongly opposed the body-as-a-machine concept. He argued that life was not determined by physics and chemistry, but was the result of special biological processes. Although the chemical events in the living body might appear similar to those discovered in the laboratory, this similarity is only superficial. Stahl's fundamental idea was that the body was merely the temporary house for the immortal soul, or anima. By itself, the body tended toward dissolution, but this was prevented by the anima. This anima was responsible for making the local changes and adjustments in the body necessary for its preservation. At the core, therefore, physiological processes are governed by a soul. Stahl was a champion of vitalism, the position that living forms have unique intrinsic properties that cannot ulti-

Fig. 14.4. Nonconformity had its price. Mob storming Priestley's house, July 14, 1791.

mately be understood, since, by their nature, they are not amenable to scientific explanation. Sympathy with this view by religious-minded scientists (who were the vast majority at that time) perhaps lent credibility to Stahl's erroneous, but highly influential phlogiston theory, discussed later.

The vitalists' position, which was that some biological phenomena do not yield to a physical or chemical explanation, and that there is a "life force" involved that cannot be further reduced, has a long and enduring history. The production of organic chemicals, biological heat, vision, and bird navigation were all once thought to be incapable of scientific explanation. But, as we are seeing, and will see further, the story of vitalism is a progressive reduction in the number of "cannot be reduced further" phenomenon as physiology, neurophysiology, and molecular biochemistry advance.

15 The Enlightenment and Rational Biology

CONTENTS

Strongly influenced by the success and prestige of Newtonian physics, eighteenth century biological models reflected balance in nature, epitomized by the concept of natural law in which the components of the system are eternally revolving (seasonal), but never changing (developing). In the eighteenth century, an enduring dispute emerged between the reductionists—all life can essentially be reduced to the principles of chemistry and physics—and the Vitalists—living forms have unique intrinsic properties that are not amenable to scientific investigation.

RATIONAL BIOLOGY

Right up through the time of the Renaissance, scientific concepts had been very much enclosed and limited by contemporary political and religious ideologies. Finally, however, in the eighteenth century—in spite of ridicule from some poets and playwrights—the astonishing success of science in discovering the new worlds of the very large and the very small, and in providing fresh and plausible explanations of natural phenomena, increasingly influenced ideology. The philosophical mood of the eighteenth century "Enlightenment" was optimism. It was believed that, through the use of reason, which had proved so wonderfully powerful in science, humans could understand the workings of society and the mind, and that "rational thought" could illuminate the way to a society for everyone, in which the ideal human lives in the ideal society. Kant explained that this enlightenment was humanity at last coming of age, out of a long state of self-caused immaturity, which had hampered the use of reason. The motto of the Enlightenment was "Have the courage to use your own intelligence" (Rothschuh, 1973).

The concept of the balance of nature underpinned these ideas, epitomized in the Newtonian explanation of planetary motion. Newton's universal laws of motion had swept society. Indeed, the poet Alexander Pope made Newton an agent of God "Nature and Nature's laws lay hid in night: still God said *Let Newton be!* and all was light."

In Newton's universe, there is an unchanging, eternal balance between the centrifugal forces, which pull the heavenly bodies apart, and gravity, which attracts them. Once set in motion, the dance

119

of the planets and stars will continue forever, like a perpetual clock. Biological nature was similarly believed to be designed to balance the plant and animal kingdoms, and the herbivores and carnivores, the only changes being those produced by the cycle of seasons and the cycles of generation. For some philosophers, this balance of nature was an inspiration for finding the rules or laws necessary for a rational balanced natural government, a constitution of checks and balances, which protected against the tyranny of a few on the one hand and mob rule on the other.

However, the "classical" concept of natural balance allowed no room for emergent novelty: Once harmony is achieved, the arrangement is perpetual. Thomas Malthus (1766–1834) unwittingly introduced a destabilizing notion in *An Essay on the Principle of Population as it Affects the Future Improvement of Society* (1798). He pointed out that the human population has an innate capacity to increase in numbers at a rate greater than the food supply. There is, therefore, he gloomily observed, a natural balance in the distribution of wealth and resources, produced by the solid check of a limited food supply, and man (in general) will forever be on the verge of starvation. This is a natural law that makes poverty and starvation the lot of most of mankind, and to oppose it by charity is to act against natural providence, a warning readily accepted by many of the new industrialists. But, as Darwin was later to notice, the model of an unchanging balance of nature had received a fatal blow: If there are more animals than food, those who are better equipped will survive to reproduce; those who are not, will perish, which will change the composition of the next generation, and the next and the next.

THE EMERGENCE OF PHYSIOLOGY

The ever-increasing growth in the scope of scientific activity in the eighteenth century resulted in the establishment of particular scientific disciplines, each with its own area of expertise. Two major divisions appeared: natural philosophy, which deals with universals, describes scientific

laws that apply now, in the past and future, here and throughout the whole universe, and is expressed in mathematical equations; and natural history, which accounts for and explains particular natural phenomena in descriptive terms.

As the biological sciences became more specialized, physiology emerged as a subject worthy of independent study. In 1747, Albrecht von Haller (1707–1777), an anatomist, physiologist, and an accomplished poet, published what has been called the first textbook of physiology, his eight-volume *Physiological Elements of the Human Body.* He described the work of a large number of scientists in detail, and reviewed their conclusions, rather as in a modern physiological review. He imaginatively defined physiology as *animata anatome* (vitalized anatomy). Haller viewed physiology as an empirical science: The organism is a machine whose internal processes are kept in motion by a series of forces which could only be known by their effects. He stressed the importance of physics and chemistry in providing explanations.

Haller was particularly interested in describing the essential feature of life, and he performed experiments on hundreds of animals to find out what parts were irritable and what parts sensible. The irritable parts are those that contract when touched, the sensible parts conveyed a message to the mind. He found that the stimulus of pinching, pricking, or certain chemical agents evoked contraction in muscle and intestine, but not in liver or kidney. He tested the sensibility of various brain structures by pinching with forceps, and by chemical irritation and found that the cortex was completely insensitive, but that stimulating white matter produced expressions of pain and the animal attempted to escape. As he himself admitted, many of these experiments were extremely cruel.

Haller championed the doctrine of "irritability" as a phenomenon peculiar to life. Irritability is "widely present not only in the animal but also in the vegetable kingdom, a contractile force by which the elements of fibres are brought nearer to each other." More importantly, he proposed that an organ's response to an external stimulus was the result of its being prepared to react in some specific way, and was not simply the cause-effect of the stimulus expressed. An analogy was the spark triggering a gunpowder explosion!

This concept—that biological systems were functionally organized, and not simply consequences of physics and chemistry—was to prove extremely fruitful. Haller's books were very popular, and were still in use in the nineteenth century. Indeed, according to Lusk (1933), Haller's *First Lines of Physiology* was the only physiology textbook used by American medical students between 1787–1815.

The subsequent exponential expansion in physiology was so great that we can really only cover a few topics in this short book. For this period, we have selected digestion, respiration, bioelectricity, and reproduction, to illustrate the maturation of physiology in the eighteenth century.

Digestion

As we discussed in Chapter 12, there was a heated dispute over the agent that caused food to be changed into absorbable chyle. The essence of the argument was: Is digestion merely the result of mechanical crushing of the food (trituration); is it caused by some chemical process, or is it simply a dissolving?

The iatrophysicists, who claimed that all biological phenomena could be explained by physics, considered digestion as a straightforward mechanical grinding of food into smaller and smaller particles. For example, Boerhaave argued that bones were not digested, but crushed to powder.

They were opposed by the iatrochemists, who believed that all biological phenomena were ultimately chemical. Helmont and Sylvius, in particular, contended that the digestion in the stomach was a chemical process brought about by fermentation or putrefaction (which was thought to be one of the modes of fermentation). This fermentation was thought to be caused by secreted juices, such as saliva from the mouth and fluid discharges from the walls of the stomach. However, for Haller, gastric juice was not a ferment but a dissolver.

The breaking of this deadlock was brought about by the elaboration of more and more sophisticated experiments to investigate digestion. René de Réaumur (1683–1757), who improved the thermometer and established a temperature scale (the Reaumur scale), asked the question: Is the "chymification" of the food in the stomach the simple result of mechanical trituration, or is it brought about by chemical putrefaction, or by a dissolving property of the gastric juices? To answer this question, he fed a kite small metal tubes that contained meat. The tubes were open at both ends, and the meat was secured by a fine wire grating. The tube was retrieved when it was regurgitated (it is the habit of such birds to throw up what they cannot digest) (Foster, 1901).

Réaumur noted, in his treatise on the *Digestion of Birds* (1752), that, although the meat had been partially dissolved, he could see no sign of putrefaction. Repeating the experiments with different contents in the tubes, he found that, although small pieces of bones dissolved in the stomach, vegetable grains were little altered. On closer observation, he noted that sometimes the regurgitated tubes were filled with a yellowish fluid whose taste was salty and bitter, and which he believed was the agent of dissolution. He cleverly obtained a sample of this dissolving fluid by placing pieces of dry sponge in his tube, and found that the liquid he got by squeezing the retrieved sponge was acid, "which turned blue paper red." Unfortunately, the kite died (possibly of starvation), and the bird experiments were discontinued, but Reaumur had invented a wholly new method for investigating digestion.

The next step was taken by Lazzaro Spallanzani (1729–1799), one of the most inventive experimental scientists of all time (he made important discoveries in sterilization, reproduction, respiration, bioelectricity, and bat echolocation, some of which will be mentioned later). He confirmed and greatly extended Rémeaur's findings, using modified gastric tubes on a wide variety of animal species, including fishes, frogs, cats, dogs, and sheep, and retrieving them from different parts of the animal's intestines. He also experimented on himself by

122 | The Rise of Experimental Biology

	The several trials	The quantities of blood let out in wine measure		The several heights of the blood after these evacuations	
		Quarts	Pints	Feet	Inches
*These five ounces lost in preparing the artery	1	0	*5 ounces	8	3
	2	1	0	7	8
	3	2		7	2
	4	3		6	6½
	5	4		6	10½
	6	5		6	½
	7	6		5	5½
By this time there is a pint lost	8	7		4	8
in making the several trials which	9	8		3	3
is not allowed for in this table.	10	8	1	3	7½
	11	9	0	3	10
	12	9	1	3	6½
	13	10	0	3	9½
	14	10	1	4	3½
	15	11	0	3	8
	16	11	1	3	10½
	17	12	0	3	9
	18	12	1	3	7½
	19	13	0	3	2
	20	13	1	4	
	21	14	0	3	9
	22	14	1	3	3
	23	15	0	3	4½
	24	15	1	3	1
	25	16	0	2	4

Fig. 15.1. Hales' data on the amount of blood that "issued on severing the carotid arteries of 25 horses," and the height it rose to in connecting tubes. (From *Haemastaticks*.)

swallowing small linen bags and perforated wooden tubes containing meat, bread, and so on which he rescued for careful examination, after they had been voided via the anus. He studied human gastric juice, got by making himself vomit on an empty stomach before breakfast, and tested the effects of gastric juice on various foods, kept warm by keeping the container under his armpit for 23 days, or placing it in a stove (Foster, 1901).

From these and many other experiments, Spallanzani concluded that the gastric juice of all animals had the power to dissolve food into "chyme," and that this action was more effective if solid materials, such as bones or grains, was first ground to a powder, or if solid vegetable matter was masticated into a pulp with saliva. Trituration, therefore, was a preparation for dissolution. He declared that digestion was not caused by putrefaction, because meat placed in warm water readily putrefied; meat in warm gastric juice for the same period remained sweet (a control experiment). But

he also concluded that acid played no part in digestion, since he had found that his fasting gastric juice was not acid.

The error in this observation was soon corrected by Bassiano Carminati (1750–1830), in 1785 who showed that, in carnivores at least, the gastric juice is neutral when the animal is starving, and becomes strongly acid shortly after it is fed. But this finding was ignored, and it was not until 1824 that William Prout found free hydrochloric acid in the stomach, and it began to be realized that what takes place in the stomach is only the first of a series of changes that occur during the passage of food down the gut. Spallanzani had found what he called a "powerful solvent" in the stomach, but its nature was elusive. He believed that it did not act as an acid nor did it produce fermentation. Later, the Scottish surgeon John Hunter (1728–1793) noted that the stomach self-digested following death, and argued that this showed a vitalistic "living principle" was involved in gastric digestion.

The feveral Animals.	Quantities of Blood = to the Weight of the Animal in what Time.	How much in a Minute.	Weight of the Blood fuf-tain'd by the left Ventricle contract-ing.	Num-ber of Pulfes in a Mi-nute.	Area of the tranf-verfe Sec-tion of de-fcending Aorta.	Area of the tranf. Sec-tion of af-cending Aorta.
	Minutes	Pounds	Pounds		Square Inches	Square Inches
Man	36.3 18.15	4.37 8.74	51.5	75		
Horfe 3d	60	13.75	113.22	36	0.677	0.369
Ox	88	18.14		38	0.912	0.85 Ri. left
Sheep	20	4.593	35.52	65	0.094 0.383	0.07 0.012 0.246 Ri. left
Dog 1	11.9	434	33.61	97	0.106	0.041 0.034
2	6.48	3.7			0.102	0.031 0.009
3	7.8	2.3	19.8		0.07	0.022 0.009
4	6.2	1.85	11.1		0.061	0.015 0.007
					0.119	0.7 0.031
					0.125	0.062 0.031
7	6.56	4.19			0.109	0.053 0.032

Fig. 15.2. Hales' calculations on the weight of blood ejected by the left ventricle contracting in several animals (including man), and the cross-sectional areas of their respective "ascending" and "descending" aortas. (From *Haemastaticks*.)

Respiration, Blood, and Animal Heat

The new quantitative technologies of gas pumps, volumetric methods for gas analyses, and improved thermometers facilitated a fresh approach to the problems of respiration and animal heat.

Stephen Hales (1677–1761) a clergyman with the perpetual curacy of Teddington, showed an outstanding ability for quantitative experimentation. Indeed, he has been called a veritable "Newtonian physiologist" (Guerlac, 1977). Hales was obsessed with numbering, weighing, and measuring biological phenomena, and gave the name "staticks" to this approach. His two famous books were correspondingly entitled *Haemastaticks* (1733) and *Vegetable Staticks* (1727).

The impetus for Hales's famous investigations into the dynamics of blood flow rose from a dissatisfaction with Borrelli's claim that muscular motion is caused by the force of blood swelling the muscle (Hall, 1987). In order to investigate this problem, he needed to accurately measure blood pressure and "it occurred to me that by fixing tubes in the arteries of live animals, I might find pretty nearly, whether the blood, by its mere hydraulic energy, could have a sufficient force, by dilating the fibres of the acting muscles, and thereby shortening their lengths to produce the great effects of muscular motion" (Lewis, 1994).

Hales launched a series of experiments tapping the carotid arteries and jugular veins of dogs, horses, deer, cows, and sheep. In his most classic experiment described in *Haemastaticks*, he "caused a mare to be tied down alive" and inserted a brass cannula, with a one-seventh-inch bore, into the exposed left carotid. The windpipe of a goose was connected to the cannula to give it pliancy, and a long glass tube of nearly the same diameter was attached to the other end of the wind pipe and put into a vertical, position. Hales then untied the ligature on the artery and, as the horse died, noted that the blood quickly rose in the tube to a height of 9

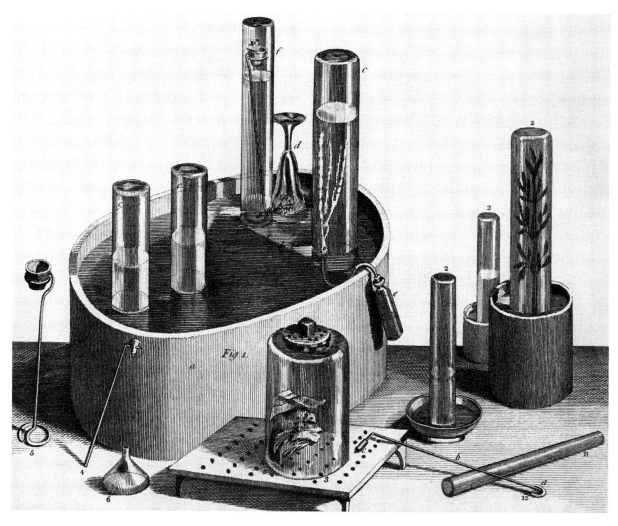

Fig. 15.3. Priestley's gasometer. He used mercury as the confining liquid (**a**). Mice are seen in the broad glass cylinder (**3**). Phlogiston was absorbed from the air on heating red mercuric oxide in a confined space (**f**). (From Wellcome Trust.)

feet 6 inches perpendicular above the left ventricle of the heart, and that it would rise and fall with the pulse, sometimes as much as 12–14 inches. These experiments were carried out many times (Fig. 15.1). He measured the weight of blood in the body and heart; estimated the surface area and volume of the left ventricle using paper covering and a casting made from bees wax; and calculated the speed of the blood in the veins and other blood vessels (Fig. 15.2).

From these data, he was able get an estimate of the blood pressure and the force produced by the heart. He studied peripheral resistance, by comparing the rate of blood flow through an isolated organ after injecting various substances, such as brandy

and various salines. The large changes in resistance that were produced were caused, he thought, by changes in the diameter of the capillaries, or in modern terms, vasodilation and vasoconstriction. Hales concluded that the force of blood in the capillaries could only be very small, and would be insufficient to produce "so great an effect, as that of muscular motion." He had pioneered the science of hemodynamics.

Hales argued that the secret to the function of the blood, lay in the newly discovered red globules, which serve to hold the heat. The color of the blood resulted from strong agitation and friction. This can be seen by agitating blood in a closed glass

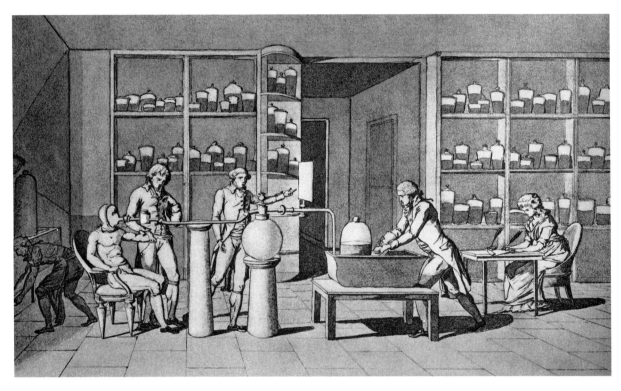

Fig. 15.4. Lavoisier's respirometer. Madame Lavoisier, who was an active participant in the experiments, is taking notes.

vessel, where it becomes quite florid. Blood "principally acquires its warmth, by the brisk agitation it there (in the lungs) undergoes." In agreement with ancient ideas, he held that respiration serves to cool the blood; as evidence: The modern thermometer confirms that cool air is inspired and warm air expired.

Although Hales made many original and important contributions to plant and animal respiratory physiology, he thought that his most practical accomplishment was his invention of ventilation systems for use in ships or prisons. And, as a person of his time, he justified his scientific labors in religious terms. As he says in *Haemastaticks,* it was a duty to discover and wonder at the wisdom and goodness of God: "the All-Wise framer illustrates the wisdom of the divine Architect."

Respiration

It had been clear, even to the ancients, that respiration was necessary for life. Hooke had demonstrated that life was not dependent on the

mechanism of breathing by itself, but on some quality of the air. But what was this quality? In his *Elements of Chemistry* (1735), Boerhaave, who suspected that it might be related to the loss of elacticity in the air, asked if there exists "an air of special virtue for the lives of animals and plants," and declared, "Many chemists have announced the existence of a vital element in the air, but they have never told what it is or how it acts. Happy the man who discovers it!" (quoted in Lusk, 1933).

It was generally accepted that air was necessary for combustion, and the dominant theory of combustion at that time was the phlogiston theory of Stahl, who proposed that all combustible materials held the element phlogiston, and that, during combustion, phlogiston was released to "mingle with the air." As air accumulated, phlogiston combustion was depressed, but, as air became dephlogisticated, so combustion was encouraged. A material that had lost phlogiston in combustion could only become combustible again by retrieving phlogiston back out of the air. For the next 40 years or more, many scientists were involved in a fruitless, and as we shall see, diversionary search for this elusive substance, one that had the properties of being invisible and weightless, or even having negative weight.

Joseph Priestley (1733–1804) made important advances in understanding how air supported life, through the use of the latest improved air pumps, and by collecting gases under mercury (Fig. 15.3). Priestley was a Unitarian clergyman of distinctly unorthodox political views. A friend of Benjamin Franklin, he sympathized with the American and French revolutions. But it was a time of little tolerance. After his house in Birmingham was burned down by a mob (Fig. 14.4), he found it expedient to move to London, then, later, to flee to the United States.

Priestley was prolific, writing about 150 books, mostly theological (he thought theology much more important than science). But it is his works on air and other gases for which he is remembered. In his most famous book, the three-volume *Experiments and Observations on Different Kinds of Air*, he records a most important discovery: "I have been so happy as by accident to have hit upon a method of restoring air which has been injured by the burning of candles" (Foster, 1901). As he further noted, "this restoration of the air, I found depended on the vegetating state of the plant". He had discovered that he could revitalize the air in a glass chamber that had been exhausted by animals and flames, by the common method of placing a sprig of mint in the inverted jar. He had made the astonishing discovery that vegetation restores air's life-giving properties. Unfortunately, and to history's great frustration, he pursued a phlogiston explanation of this phenomenon: He posited that the burning candle filled the chamber space with phlogiston, and the growing plant absorbed it!

Priestley's most original and significant contribution was the discovery that heating mercuric oxide within a chamber, with a burning-glass (magnifying glass) held outside (to avoid contamination), produced a gas that not only caused a candle to burn with "more splendour and heat," but allowed a mouse to survive 4–5 times longer than in a chamber of ordinary air. He had thus discovered oxygen, but could not see it. As a convinced phlogistonist, he speculated that, when mercuric oxide was heated, it took up all traces of phlogiston from the air to make "pure dephlogisticated air."

A respiratory gas that would not support life was discovered by Priestley's contemporary, Joseph Black (1728–1799), who held chairs at Glasgow and Edinburgh Universities, and was an advocate of the use of the balance and thermometer in experimentation. He found that, if he heated chalk, a nonflammable gas was produced, which he called "fixed air," and noted that it "is deadly to all animals." He had rediscovered, although in much more quantitative terms, Helmont's gas sylvestre, of 100 years earlier. Through careful and accurate weighings, he noted that the amount of weight lost, when a measured quantity of chalk was burned into quicklime (calcium oxide), was fully regained when the chalk was regenerated by boiling the quicklime in mild alkali (potassium carbonate). In contemporary terms, he had produced the following reaction, and had founded quantitative chemistry:

$$CaCO_3 <—> CaO + CO_2$$

He discovered that fixed air (carbon dioxide) could be detected by bubbling the gas into a solution of lime water (calcium hydroxide), whereby a white, milky precipitate of lime (calcium carbonate) appeared. By this means, it could be simply demonstrated that alcoholic fermentation, burning charcoal, and breathing animals all gave off fixed air, and the amount given off could be accurately quantified. It is recorded that, in 1764, he collected air from a duct in the ceiling of a church in Glasgow, where 500 people had gathered for over 10 hours in prayer—a dour comment on the habits of the times.

In his thermal studies Black for the first time established a clear distinction between temperature and heat, and demonstrated the different kinds

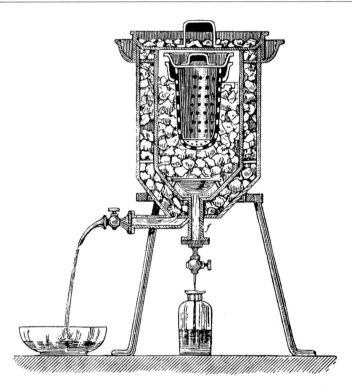

Fig. 15.5. The calorimeter of Lavoisier and Laplace. An animal is placed in the inner chamber, and a quantitative estimate of animal heat is obtained by measuring the amount of ice melted.

of heat (i.e., heat capacity, specific heat, and latent heat), thereby providing the intellectual tools to at last tackle the old problem of innate heat.

In 1777, Adair Crawford (1748–1795), Black's pupil, published a theory of heat that connected animal heat and respiration in a novel way. Using a self-designed calorimeter that measured changes in water temperature, he found that "the quantity of heat produced when a given quantity of pure air (oxygen) is altered by the respiration of an animal is nearly equal to that which is produced when the same quantity of air is altered by the combustion of wax or charcoal" (Lusk, 1933). He thought that animal heat was closer to the combustion of wax because, like respiration, burning wax also produced water, and he asserted that "a man is continually deriving as much heat from the air as is produced by the burning of a candle." This was not that unusual, considering that the resting power output of a human is about 80 watts, and that a candle might produce 100–150 watts.

But what is the nature of this combustion-respiration? The answer was provided by the famous and prolific French scientist, Antoine Lavoisier (1743–1794). As a wealthy aristocrat, he could afford the finest of instruments, and owned a balance that weighed 600 g + 5 mg, and had the most accurate gasometers available, in which gases were collected in mercury-filled tubes. He knew of Crawford's work on animal heat and Priestley's studies on respiration. He repeated Priestley's experiments, but from a different approach, and discovered that, in contradiction to the phlogiston theory, when mercuric oxide is heated, weight is lost and a gas given off. This gas not only supports life and combustion, but it is one-fifth part of normal air. He showed that this "eminently respirable air" was converted into Black's "fixed air" by both combustion and respiration, and in 1777 he called the gas "oxygen." With his wife as his main collaborator, he conducted a series of experiments measuring and comparing the rates of oxygen con-

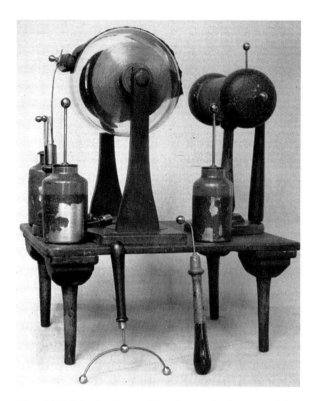

Fig. 15.6. Priestley's machine for producing electricity. (From Science Museum, London.)

sumption and carbon dioxide produced, and showed that the intensity of oxygen consumption was dependant on food intake, the environmental temperature, and mechanical work (Fig. 15.4).

With the help of the mathematician Pierre Simon de Laplace (1749–1827), Lavoisier designed and had built an ingenious apparatus, an ice calorimeter, for directly measuring metabolic heat. The rate of body heat production was accurately measured by his ice calorimeter, as the amount of ice melted by an immersed animal (often a guinea pig), in a fixed amount of time (Fig. 15.5). They also designed a combustion chamber to demonstrate that the burning of hydrogen (recently discovered by Cavendish) and oxygen produced water, and to prove that Black's fixed air is a compound of carbon and oxygen. These studies drove him to the great conclusion that the respiration results in oxygen from the air combining with carbon and hydrogen from the body to produce water, fixed air

(carbon dioxide), and heat, the carbon and hydrogen being derived from food.

By measuring the corresponding quantities of oxygen consumed and fixed air produced, Lavoisier was able to accurately compare the processes involved in burning a candle and animal respiration. He concluded that animal heat is the heat produced by oxidation, and that it is essentially like the slow burning of a candle. Lavoisier believed that oxidation of the body carbon and hydrogen occurred mostly in the lungs, which therefore must be the site of animal heat. In this scheme, the blood's role is the old one, to distribute this heat, but the lung is the source of heat, and not the traditional heart.

Joseph Louis Lagrange (1736–1813), a French mathematician and physicist, disagreed with Lavoisier over the site of oxidation. If it was the lungs, then they should be warmer than the rest of the body, so hot that "one would have reason to fear they would be destroyed" and this was not so. He proposed instead that oxidation occurs in the blood as it flows throughout the body; otherwise, why is the body temperature uniform?

Spallanzani, inspired by Lavoisier's discoveries, launched a long series of experiments on the respiration of both vertebrates and invertebrates, under a variety of conditions, including the effects of temperature and hibernation. He showed that tissues, such as stomach, liver, and the ovaries of fishes, absorbed oxygen and gave off carbon dioxide. Oxidation, therefore, occurred at the tissue level, even in animals without lungs. But, yet again, he was before his time, and his findings were ignored. Lavoisier's view that oxidation takes place in the lungs held sway for almost the next half-century. The question was not really resolved until 1837, when Heinrich Gustav Magnus (1802–1870) was able to show, with the use of a mercurial vacuum pump, that arterial blood had a higher oxygen content than did venous blood, and that venous had a higher amount of carbon dioxide than arterial, i.e., oxygen-carbon dioxide exchange was not made in the lungs or the blood, but in the tissues.

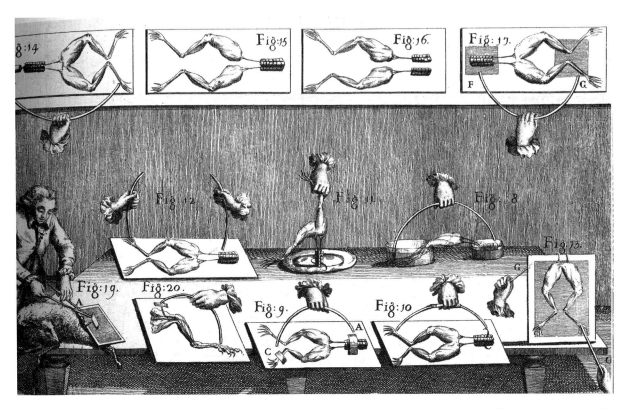

Fig. 15.7. Galvani's demonstration of animal electricity. In Fig. 17 (*inset*), part of the spinal cord is lying on a silver dish (**F**). The frog legs are resting on a copper dish (**G**). A wire completes the circuit.

Ironically, while Priestley fled from England to the freedom of the United States following the violent protest against his support for the French Revolution, Lavoisier was arrested in France by the revolutionary authorities. He was condemned by the revolutionary tribunal for holding the hated position of farmer-general of the revenue, who had the authority to enforce and collect at a profit government taxes in their private tax farms, and was guillotined with 27 others on May 8, 1794.

Bioelectricity

Since ancient times, knowledge of electrical phenomena came from four sources: lightning; rubbed amber (Greek electron), which attracts such materials as paper and feathers; the lodestone, or magneta stone, which attracts iron (this is mentioned in Homer, and the magnetic needle was used as a compass in medieval Europe, and much earlier in China); and the electric ray *Torpedo mamorata*

and the electric catfish *Malapterus electricus* (their shock was used by Greeks and others to treat headaches and epilepsy).

However, until the eighteenth century there was no real awareness of any connection between these phenomena. With the development of frictional electric machines, electricity became a popular topic for serious and amateur inquiry (Fig. 15.6). Considerable charges could be built up in huge Leyden jars (a sort of early capacitor invented around 1750 by Dutch physicist Pieter Von Musschenbroek and named after his city). These charges were sufficient to kill small animals on discharge, and to jolt a holding-hands chain of 180 of Louis XIV's guards or 700 monks from the convent de Paris (Katz and Hellerstein, 1982).

Static electricity machines also inspired more serious studies. Several scientists commented on the obvious visual similarity between lightning and the electric spark produced on discharge. Their

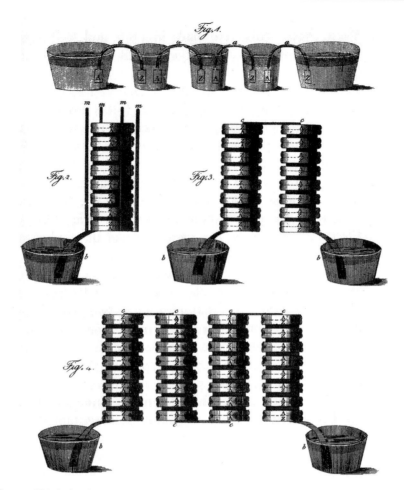

Fig. 15.8. Four variants of Volta's electrical pile or battery. Moist paper separates silver (**a**) and zinc (**z**) plates.

identity was established by Benjamin Franklin, who bravely charged a Leyden jar with lightning. The similarity between the shock of the electric fish and that of the "electrical machine" was also noted in Oliver Goldsmith's *Animated Nature* (1774): "the instant it (the electric ray) is touched, it numbs not only the hands and arm, but also sometimes the whole body. The shock received, by all accounts, most resembles the stroke of an electrical machine; sudden, tingling, and painful." In 1773, John Walsh suggested, in a report to the Royal Society, that the shock from the torpedo fish was caused by the release of compressed electric fluid (Katz and Hellerstein, 1982). It was thought that these "electrical fluids" had some association with the "nervous fluid," which was secreted by the brain and conducted to the muscles via the nerves where it became an agent of contraction (Mauro, 1969).

The watershed observations on animal electricity were made by Luigi Galvani (1737–1798). In his 1791 treatise, *Commentary on the Effects of Electricity on Muscular Motion*, he noted, "I placed the (dissected) frog on the same table as an electrical machine ... at a considerable distance from the machine's conductor ... When one of my assistants by chance lightly applied the point of a scalpel to the inner crural nerves ... suddenly all the muscles of the limbs were seen to contract."

The next experiment was even more fundamental. Galvani hung frogs' legs, which had been fastened by brass hooks in their spinal cords, to an iron railing, and noted that they twitched when the feet touched the balcony, but he thought that this

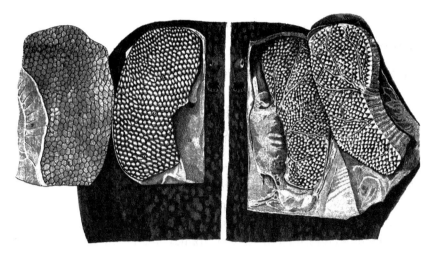

Fig. 15.9. Dissection of torpedo fish by John Hunter, to reveal the electric organs. The left illustration shows the upper surface of the organs; the nerves are shown in the right.

response might result from electricity in the atmosphere. He therefore repeated the experiment indoors. The muscle was placed on an iron plate, and, as soon as he touched the brass hook to the plate, "Behold! The same contractions and movements occurred as before." Galvani believed he had discovered "animal electricity," and that the metals had provided an arc for conduction from the nerve to the muscle, discharging the accumulated electricity in the muscle, rather like a Leyden jar discharging (Fig. 15.7)(Piccolino, 1997). His concept was that electrical fluid, secreted in the brain, passed down the nerve, and was collected in the muscle. Galvani noted that different combinations of metals produced different degrees of contraction, but thought this of no great importance. In his view, the external arc played a passive role in transferring electricity from nerve. This idea was reinforced by his observation that nonconductors, such as wood and glass, produced no muscular contraction. He sent one copy of his privately printed book, inscribed from the author, to his compatriot, Count Alessandro Volta (Piccolino, 1997).

Volta (1745–1827), professor of physics at the University of Pavia, was particularly intrigued by Galvani's finding that the vigor of contractions depended on the kind of metal used to form the external arc, and, disagreeing with Galvani, came to the conclusion that the electrical fluid did not arise from the organism, but from the contact between dissimilar metals. It was not a case of "animal electricity," but of "metallic electricity." Galvani had unknowingly discovered a new way of making electricity, not a new kind of electricity. Volta vigorously attacked the concept of animal electricity, and a long, sometimes acrimonious, debate ensued, with partisans on both sides. Sir Joseph Banks, President of the Royal Society, lauded Volta for his "infinite acuteness of judgement," while the celebrated polymath, Alexander von Humboldt (1769–1859), demonstrated that muscle contraction could be produced without the bimetallic arc, by cutting and reconnecting the crural nerve, or by peeling off muscle from the sciatic nerve. However, Galvani became disheartened over the bitter controversy, and withdrew from science.

Volta pursued his ideas, and, in 1800, he reported to the Royal Society the invention of an apparatus "which will, no doubt, astonish you." It consisted of a multilayered sandwich of alternating thin disks of silver and zinc, separated by pieces of pasteboard soaked in salt water, and, unlike the friction machines, this "pile" produced a steady electrical current, instead of a single discharge. Volta had

Fig. 15.10. Electrotherapy, early twentieth century. The subject looks suitably subdued.

These novel biological effects, produced by the new "animal electricity," stimulated great popular interest. Most dramatically, it was universally accepted that the basic distinction between death and life was that, to be dead was to be immobile, to be alive was to have movement. Was electricity a means to restore the dead to life? To many it seemed that it very much might be so. There followed many curious and bizarre experiments to bring dead animals back to life through electric shock, and even executed humans were zapped by eager experimenters (since the procedure was morally and legally dicey, much was done in secret). For example, a Dr. Aldini (quoted in Gatrell, 1994) described his treatment of a murderer, Foster, who was executed in Newgate in 1803. After shoving one electrode up his rectum and another in his ear, Aldini connected the battery, and noted how Foster's muscles got "horribly contorted and the left eye actually opened," which gave the corpse the "appearance of reanimation."

Mary Shelley's novel, *Frankenstein* (1818), is premised on this enthusiasm to defeat death through life-restoring electricity. Elaborate quack therapies mushroomed around "electrical medicine" and Mesmer's "animal magnetism," involving such treatments as placing patients in tubs filled with iron filings to restore any magnetic deficiency or wiring the patient to discharge healing electric sparks (Sutton, 1981; Fig. 15.10). Similar scams remained popular at the turn of the twentieth century (Fig. 15.11), and are still profitable today.

Reproduction

We saw earlier that Aristotle considered that the female provided the material for the developing fetus (in some form of menstrual blood). Like Aristotle, Harvey was also an "ovist" with a preformationist view of embryology, believed that the shaping power itself was contained in the egg. The organs of the embryo were thought to arise by a process of new formation, called "epigenesis," with development proceeding from undifferentiated beginnings to the complex newborn. Harvey's analogy was to compare the fetus to a house or

invented the battery (Fig. 15.8). Rather ironically, considering his untrammeled attack on Galvani's "animal electricity," he believed that he had constructed "in its form ... the natural electric organ of the torpedo fish" (Fig. 15.9), and named his invention the "artificial electric organ." Volta eagerly investigated the biological effects of his voltaic pile, as it became better known. A frog leg touched by this electrical machine would continue to contract until it was exhausted; electric terminals placed on muscle caused involuntary contractions, and, placed on the tongue, produced a sensation of taste. Volta reported a self-administered experiment in which he pressed one terminal against a moistened eyelid, while the other touched the tongue: "A beautiful flash" was produced. The science of electrophysiology was underway.

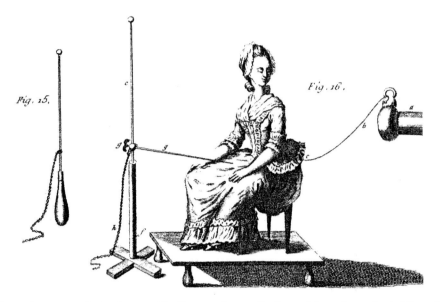

Fig. 15.11. Eighteenth century electrotherapy. Healing sparks are discharged from the patient. (From P. J. C. Mauduyt de laVerenne, 1784.)

ship whose frame is laid down first. Charles Bonnet (1720–1793) compared epigenesis to crystal growth. The embryo begins as an exceedingly fine net on the surface of the yolk, fertilization makes part of it beat, and this becomes the heart, which, sending blood into all the vessels, expands the net. This net then catches up food particles in its pores (Needham, 1959).

But in the seventeenth century, epigenesis was mostly overthrown by the discoveries of Malphigi and Leeuwenhoek, who saw, under the microscope, that the embryo was not simply an undifferentiated object. In the rival, preformation theory, the development of the embryo is essentially an enlargement or opening of the preexisting organs in the "germ" of the species. At that time, the term "evolution" referred to the "unfolding" of the tiny but complete baby contained in the germ. By the 1720s, the theory of preformation was thoroughly established.

But where was this minute creature held, in the egg or in the sperm? Bonnet reasoned that, for a machine to work properly, all parts are necessary; therefore, for the developing animal, all parts must be provided at the beginning, in the germ located in

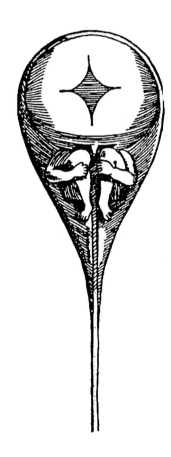

Fig. 15.12. Hartsoeker's famous drawing of a human spermatozoon (1694).

the egg. Before fecundation, the egg contains an excessively minute but complete baby, which "evolves" during development. Bonnet accepted the logical, but difficult, consequences of this position: that every female contains within her the germs of all children who would originate from her, and their children's children, from generation to generation. The purpose of the semen was merely to initiate growth. There were many preformation "sightings." For example, J. B. du Hammel claimed that he could actually see the chick embryo in the egg before fertilization, and Jacobaeus said the same for frog eggs (Needham, 1959). Even more extravagantly, the anatomist, Theodor Kerckring, (1640–1693) reported, in *Observantiones a themedicae* (1672), seeing bird-like eggs in a female killed in flagrante delicto, and human embryos, 3–4 days old, with traces of head and organs, in virgins (Wilson, 1995).

In opposition, the animaculists, who included Leeuwenhoek, Leibnitz, and Carl Ludwig, believed that the preformed germ resided in the sperm that Leeuwenhoek had discovered in the male ejaculate. The sperm were thought to be like seeds (semen) to be planted in the fertile soil of the mother's womb. Looking hard in the microscope, many imagined that they could actually see preformed infants in the spermatozoa. For example, Gautier saw minute forms of men, complete with arms, heads, and legs, inside human spermatozoa, and a microscopic horse in the semen of a horse. In 1694, Nicholaas Hartsoeker (1656–1725) provided a famous illustration of how the human fetus might be enclosed in the spermatozoon (Fig. 15.12).

As related by Needham (1959), the animalculists had a major conceptual problem to deal with: How could such a vast number of preformed sperm babies be wasted for just one conception? Surely, nature could not be so prodigal. James Cooke provided an ingenious solution, published in 1762. He proposed that all the spermatozoa, "except that single one that is then conceived, evaporate away, and return back to the Atmosphere again, into the open air," where they "live in a latent life, in an insensible or dormant state, like Swallows in the Winter … till they are received afresh into some other Male body of the proper kind, to be again set on Motion, and ejected in Coition as before, to run a fresh chance for a lucky Conception" (Needham, 1959). The idea of the air bearing innumerable invisible human spermatozoa shades of Burton's spirit world, received wide attention, but was ripe for satire. In 1750, Sir John Hill addressed the Royal Society concerning a machine he had invented for catching the spermatozoa carried on the West wind. He erected a "kind of Trap to intercept the floating Animalculae … electrified according to the nicest Laws of Electricity … applying my best Miscroscope, plainly discern them to be little Men and Women, exact in all their Lineaments and Limbs, and ready to offer themselves little Candidates for Life," whenever they should be "conveyed down into the Vessels of Generation" (Needham, 1959).

The ingenious Spallanzani made some important discoveries on the role of semen, through a series of elegant experiments. It was thought, even by Harvey, that semen stimulated, or started, the development process, through the action of an incorporeal fertilizing power called the "aura seminalis," which Harvey likened to the invisible power of the magnet. Although Spallanzani also believed that the egg contained the preformed individual, and that the spermatozoa were "spermic worms" that lived in seminal fluid, he wanted to determine more particularly the role of semen in the process. On experiments with frog eggs, Spallanzani noted that, although eggs that had been discharged, and had come into contact with the semen, developed, eggs dissected from the same female's body did not. Did semen play a role in at least initiating development?

Spallanzani designed special tightfitting taffeta pants for male frogs, and found that none of the eggs that were discharged when these males copulated developed. However, when he mixed the eggs with semen from the males' trousers, normal development took place. He then "painted" semen on eggs, which then developed. He had proved that fertilization is not the effect of the incorporeal aura seminalis, but is caused by the "sensible part of the seed." He had discovered a method of artificial fertilization. His triumph was the artificial insemination of a bitch, which gave him, he said, more intellectual satisfaction than any other experiment he had ever done. Bonnet wrote to Spallanzani in 1781, "I do not know but one day what you have discovered may be applied to the human species to ends we little think of and with no light consequences" (quoted in Manger, 1979).

Spallanzani had proven, on precise experimental grounds, that semen was a material agent in fertilization, though he himself did not believe that it contributed to the preformed individual housed in the egg. Indeed, the matter was only settled in the nineteenth century. The mammalian ovum was not discovered until 1827, when it was described by Karl von Baer (1792–1876) in his book *On the Origin of the Mammalian and Human Ovum*, and the union of sperm and the egg nucleus was not seen until the end of nineteenth century.

During the first half of the eighteenth century, the theory of epigenesis was supported by only a few, but that group included some scientists of standing, such as Descartes and John Needham. In the second half, however, objections against preformation began to mount. What were seen to be telling arguments were made by Caspar Frederick Wolf (1737–1794), the eminent embryologist who first described the primordial kidneys, or the Wolffian ducts. Wolf believed that the chick was formed from a mass of undifferentiated little "sacs," from which the vascular system, then all other body organs, developed. Some argued that the preformation theory could not account for "monsters" (misshapen offspring). It was not comparable with the limb regeneration seen in some lower animals, and it could not explain the resemblance between young embryos of vertebrates, frogs, birds, and mammals.

But how were the individual characteristics of a baby determined? The problems of heredity and maternal and paternal contributions to the child's distinct appearance were still outstanding questions.

Even when the mother was thought to be merely a receptacle for the growing fetus, it was common knowledge that the pregnant mother's experience of strong affections could drastically affect the child in the womb. It was widely accepted that pregnant women, frightened by cows, would bear cow-like children. And why do children often look like their fathers? The explanation is that the pregnant mother, doting on the father, sends effects down the nerves to the forming child. Oliver Goldsmith, in his *History of the Earth and Animated Nature* published about 1770, tells the story of how a woman of Paris, two months pregnant, through curiosity, went to see a criminal broken alive at the wheel, but being of "a tender habit" (!), she shuddered at every blow the criminal received (a comment on contemporary mores). To the amazement of her friends, when the child came into the world, every limb was broken like those of the malefactor, and in the same place. However, he allows, complainingly, that some authors, such as Buffon, have called into question the veracity of this account. However, the concept was still alive in the 1870s. In his book *Human Physiology* (1872), C. Nichols points out that sometimes a man who is only the friend of a lady may have an (innocent) influence on the forms and features of her offspring (the effect is called "marking"). A rather convenient physiology.

16 Transition to the Nineteenth Century

Popular Science, Eccentric Science, and Romantic Physiology

CONTENTS

The transition from the eighteenth to the nineteenth century saw the last of the unbridled and sometimes eccentric science pursued by the enthusiastic virtuosi, as science was transformed into a serious and respectable profession.

HUNTER AND SPALLANZANI: THE LAST GREAT ALL-ROUNDERS

Two examples will serve to illustrate the late eighteenth-century scientist "of many parts:" John Hunter and Lazzaro Spallanzani.

John Hunter (1728–1793) was one of the last great "all-rounders" of the late eighteenth century (he has even been called "the Shakespeare of medicine" [Lasky, 1983]). Hunter made important advances in vascular surgery, and was a famous, skillful anatomist. He was curious about the power of generating heat that was peculiar to animals while they were alive. He designed his own thermometer, and took temperatures of live animals and humans subjected to extreme cold. He found that they had a power "whereby they are capable of resisting any external cold while alive, by gen-

erating within themselves a degree of heat sufficient to counteract it" (Hall, 1969b). Vital heat production did not depend on the motion of blood, because "it belongs to animals that have no circulation; besides, the nose of a dog, which is always nearly the same heat in all temperatures, is well supplied with blood" (Mendelsohn, 1964). Animal heat did not depend on the nervous system, because it is found in animals without brains. He believed that the property rested in a vital principle, "materia vitae," which was diffused through the living body, and was only found in the living (Hall, 1969b).

Hunter was particularly interested in the conditions that influence digestion (Cross, 1981). In his *Observations on Certain Parts of the Animal Oeconomy* (1786), he noted that digestion does not take place in snakes and lizards during hibernation. When live animals are placed into the stomach, the

137

Fig. 16.1. Hunter's transplant of a cock's spur into a cock's comb. (From Royal Society of Surgeons, England.) (*See* color plate appearing in the insert following p. 82.)

"dissolving powers of the stomach has no effect on them, but when dead the stomach would immediately act upon it." Indeed, he observed that after an animal dies, its own stomach quickly self-digests. Hunter provides a rather dramatic illustration of the point: "If it were possible for a man's hand, for example, to be introduced into the stomach of a living animal, and kept there for some considerable time, it would be found, that the dissolvent powers of the stomach could have no effect on them; but if the same hand were separated for the body, and introduced into the same stomach, we would find that the stomach would immediately act upon it" (*Phil. Trans. Royal Soc.*, 1772, Vol. 62, p. 449). Hunter concluded that the digestive process is also a vital phenomenon, not simply one of chemistry.

He was intrigued by avian respiration, noting for the first time that bird lungs have sacs that connect with the cavities of some bones, and that respiratory air could pass through broken bones.

Hunter investigated the growth and blood supply of deer antlers, and discovered that, if he ligatured the external carotids, the antlers first became cold, but later warm. This rewarming, he was surprised to find, resulted from the compensating enlargement of small blood vessels that restored blood supply to the antlers.

He was interested in organ transplantation, such as embedding a cock's spur in the same bird's comb (Fig. 16.1), and amused himself with rather rough and ready experiments, such as engrafting a human incisor into a wound made in a cock's comb. "I took a sound tooth from a person's head, made a pretty deep wound with a lancet into the thick part of a cock's comb and pressed the fang of the tooth into the wound and fastened it with threads passed through other parts of the comb" (Lasky, 1983). He also experimented with transplanting the testes of a cock into the belly of a hen, "where it has adhered, and has been nourished" (Lasky, 1983).

Hunter investigated infection, the processes of inflammation, and the formation of pus. In the same robust manner, in his Treatise on the Venereal Disease (1786), he described how he scarified his penis with a lancet dipped in the pus on the glans of a man infected with syphilis. He got the ailment, and his death may have been syphilitic in origin (Lasky, 1983).

We have already discussed the skillful and imaginative experiments on digestion and respiration carried out by the ingenious Spallanzani. But perhaps his cleverest and most imaginative work concerned the problem of bat navigation.

In an investigation into twilight vision Spallanzani saw that, although barn owls could easily fly around a darkened room lit by only one candle, they would hit the walls and furniture, if the candle was extinguished. Repeating the experiment with bats, he found that, in a blackened room, with no candlelight, the bats "continued to fly around as before and never struck against obstacles" (Dijkgraaf, 1960). He supposed that the bats had a special night vision that allowed them to see in what he perceived as complete darkness. Investigating further, he put opaque disks over the bats' eyes, and, to his bewilderment, saw that the bats flew normally. To settle the matter, he removed the eyes of a bat with a pair of scissors, and found that the animal flew "with the speed and sureness of an uninjured bat ... My astonishment that this animal which absolutely could see though deprived of

Fig. 16.2. An "old" museum (ca. 1599). A haphazard collection of the curious, real, and imagined. (From Ferrante Imperato, *Dell' historia naturale*, 1599.)

its eyes is inexpressible" (Dijkgraaf, 1960). The question of how the animal could "see" without the use of its eyes became known as the famous "Spallanzani's bat problem."

He investigated further. He had some tiny brass tubes constructed and fitted them into ear canals of bats. He found that, when the canals were open, the bats could fly normally. But, if he plugged the canals with pitch, the bats lost their ability to navigate, and were reluctant to fly, whether or not they had the use of their eyes. He wondered if they were sensing objects with their ears. But this question provoked ridicule. There was no conceivable scientific explanation for such a phenomenon available at the time. And, indeed, it was not until 150 years later, in the 1940s, that Galambos and Griffin

discovered that bats used high-frequency sound for echolocation (Griffin, 1959).

THE SCIENCE MUSEUM: COLLECTING AND DISPLAYING

At that time, science museums were established. These were not the haphazard collections of miscellaneous curiosities of previous ages (Fig. 16.2), but systematic displays that illustrated scientific principles.

John Hunter's anatomical museum was by far the most famous, and it was his passion (Fig. 16.3). The museum's purpose was to show comparative anatomy and illustrate physiological function, and it gradually grew to a collection of about 14,000 specimens and 500 different species. The Hunterian

Fig. 16.3. Sir John Hunter's new science museum. Displays are arranged according to a strict system of physiological classification. (From National Library of Medicine.)

Museum was divided into two main parts. First was a Physiological Series of Comparative Anatomy, which was a set of anatomical structures that had a particular use in physiology. They illustrated functions such as digestion, circulation, locomotion, and reproduction, some rather dramatically (Fig. 16.4). The second part of the museum dealt with pathological specimens and disease, and also contained other exhibits, such as an exhibition of "monsters," such as the double skull of the two-headed boy of Bengal, and human anomalies, such as an alcohol preserved six-fingered hand.

Hunter's enthusiasm for unique specimens for his museum was unbounded. For example, he funded a surgeon to go out on a Greenland whaler to bring back marine mammal parts. His most famous, or infamous, acquisition involved the Irish giant Charles Byrne (or O'Brien, as he was also called), reputedly 8 ft 4 in high. Byrne was horrified of being dissected. When he knew he was dying, in 1783, he gave strict orders to his friends for his body to be guarded day and night, and, when he was dead, to be placed in a ready-made leaden coffin and sunk at sea. Prewarned of Byrne's imminent demise, the indefatigable Hunter bribed

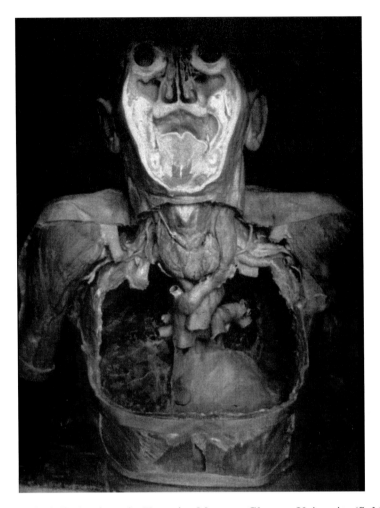

Fig. 16.4. Anatomical display from the Hunterian Museum, Glasgow University (S. Milton). (*See* color plate appearing in the insert following p. 82.)

his watchers 500 pounds sterling (an enormous amount) to smuggle the body to London. Byrne's skeleton is still on display in the Hunterian Museum in London. It indicates that he was really about 7 ft, 10 in height.

In Britain and the United States, the demand for human bodies for display and dissection severely outran the supply, and, in the spirit of the age, free enterprise provided a solution: grave robbing. The resurrectionists, as they were called, had a thriving trade in the centers of medical education in London, Glasgow, Edinburgh, and Dublin. A minor character in *A Tale of Two Cities* (1859) by Charles

Dickens, Mr. Jerry Cruncher, defines the stolen corpses of his trade, his goods, as "a branch of Scientific goods." It was a lucrative trade, the price of a body being about four pounds sterling, more than a month's salary for many, with added bonuses from burial clothes, teeth, and body fat. "I Have Made Candles of Infants Fat" was a popular song at that time (Ball, 1928). The famous were not exempt: Laurence Sterne, author of *Tristam Shandy* (completed 1766), was resurrected and dissected at Cambridge.

Private guards were hired to defend newly buried corpses, and special cages, called mortsafes,

Fig. 16.5. Mortsafes: Early nineteenth century graves armored against "resurrectionists." St. Mungo's Cathederal, Glasgow. (*See* color plate appearing in the insert following p. 82.)

were built over graves to protect the bodies (Fig. 16.5). In Scotland, the resurrectionists were careful not to take clothes, because that would have been theft, but, under Scottish law at that time, there were no property rights in a dead body, so no crime was committed. But the law of supply and demand was inexorable, and in Edinburgh the infamous Burke and Hare took the next logical step, to meet the demand for fresh dead bodies, by murdering at least 16 down-and-outs. The stealing and desecration of recently deceased loved ones naturally provoked fear and resentment. In 1788, an antiresurrectionist group raided New York Hospital dissection rooms. Horrified at what they had seen, a riot ensued. The New York Militia were called out, opened fire, and killed at least seven protesters (Ball, 1928). In Britain, the body supply problem was solved by the Warburton Anatomy Act of 1832, which authorized dissection of all unclaimed corpses or the bodies of persons who had expressed a wish to be dissected.

The richness of new ideas, speculations, and experiments promoted science as a novel popular entertainment. Public lectures that involved dissections and demonstrations became fashionable and popular and an easy target for satirists. For example, Fig. 16.6 (1802) shows the unfortunate effects produced on Sir John Coxe Hippisley from inhaling the newly discovered gas nitrous oxide. Sir Humphry Davy holds the bellows.

ROMANTIC PHYSIOLOGY

The transition from the seventeenth to the eighteenth centuries also saw the last outburst of "medieval" idealism in science. In philosophy and art, cold classical rationalism was rejected in favor of hot romanticism. Uninhibited self-expression, feeling, and intuition became the keys to true knowledge. Rousseau's noble savage, who knew by intuition how things really were, and Hegelian dialectics, which could derive by argument how things must be, were the true foundations of knowledge of nature. Romanticism swept music and art, a core concept being that nature was in some way equivalent to God (Magner, 1979).

This philosophical shift from classical rationalism had little influence on science, except in Germany, a land addicted to metaphysics. The new *Naturphilosophie,* romantic physiology,

Fig. 16.6. Rowlandson's robust illustration of a popular lecture in physiology: "An experimental lecture on the powers of air." Sir Humphry Davy holds the bellows.

flourished, dominating the period 1790–1830. It inspired the "romantic physiology" movement in what has been called by one historian "one of the most bizarre episodes in the history of physiology" (Rothschuh, 1973). Both philosophers and scientists embraced the new liberating *Naturphilosophie*, which was characterized by a disdain for empirical research and experiments. Its objective was to construct the entire material system of nature from a single all-embracing unity, all very abstract but universal.

A major exponent, Friedrich von Schelling (1783–1854), sought to provide the source from which general laws of nature were to be deduced.

For Schelling, the proper task of science was to reveal these laws, not by experiment or measuring, but by exposing the core analogies, which appeared at all levels of nature. For the philosopher, Immanuel Kant, nature was a fundamental unity of matter, process, and spirit. The most influential was the poet and philosopher Johann Wolfgang von Goethe, who made worthy studies in comparative anatomy, in which the facts were on the surface and required no experiment; he relied on the "flash of inspiration" to detect deeper truths. Goethe ambitiously wished to find the general solution to the problem of order in nature, especially living nature. And he provided an answer. Organs that superfi-

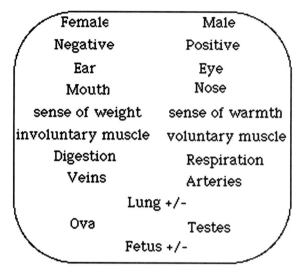

Fig. 16.7. The romantic physiologist's concept of "natural polarity" in physiology. (From T. Görres [1776–1848], adapted from figure in Rotschuh [1973].)

cially appeared to be different turned out, on closer examination, to be variants of a single, primordial plan, the "Urbild." For example, the limb bones of mammals, birds, and reptiles are all based on the same plan. Animals were variants of a basic plan, a type Urbild, such that change in one part of the organization was preserved by changes in another. This, he claimed, was seen in a reciprocal relation between the parts, for example, the possession of small feet often correlated with a big tail. With profound obscurity, he declared that the tail "may be considered as suggesting the infinity of organic existences" (Hall, 1969b).

Elaborate, all embracing "scientific" schemes were constructed in voluminous and obscure detail. The processes of living nature, for example, was thought to be organized in circles (Schubert, 1988). The major principle was "polarities," in which the properties of nature were arranged in opposite pairs, such as positive male-negative female (Fig. 16.7). Österreicher (1805–1843) championed analogy. For example, the role that the nervous system played in the circulation of the blood was seriously compared to that of the sun and the planets, and, since planetary motion around the sun was determined by the sun, so the nervous system had a directing role in the circulation of the blood.

Detailed explanations of biological phenomena were also provided. For example, the fibrin theory of body structure was a favorite among the Naturephilosophs. It was believed that chyle corpuscles turned into the white globules of the blood, which then transformed into red blood globules. The red blood globules lined up, like a string of beads (*Perlenschnur*), to form strands of fibrin. This fibrin is the basic substance of the body itself and all its organs (Mazumdar, 1974). Fibrin could be seen when blood clotted, the red clot being an aggregation of the red globules. Johannes Müller performed a simple experiment to test a critical element of this expansive theory (Müller, 1840). He filtered blood, and found that the cell-free fluid clotted. The clotting agent, fibrin, was not in the red cells. The Perlenschnur theory, and indeed romantic physiology, declined and vanished in the 1820s. The momentum and achievements of empirical physiology were far too strong.

In 1800, Karl Friedrich Burdach coined the term "biology" to denote the study of morphology, physiology, and psychology. Biological sciences had become a profession. In Germany, a member of this new profession was called *Naturwissenschaftler* and in France, *savant*. But there was no agreed English term, since the old name, "virtuoso," had fallen into disrepute and disuse. Whewell, in 1834, coined the term "scientist," from the Greek *scientia* (knowledge). In spite of some initial opposition most notably from Thomas Huxley, the term quickly caught on.

17 Consolidation of Experimental Biology

CONTENTS

The nineteenth century saw the establishment of modern physiology. Biological processes were increasingly explained in terms of physics and chemistry, and advances in physiological thinking were inspired by the new chemistry and the elaboration of experimental apparatus. For example, in his textbook Lehrbuch der physiologischen Chemie *(1857), Karl Gotthelf Lehman declared that the aim of physiological chemistry was "to discover precisely, and in their causal connections, the course of the chemical phenomena which accompany vital processes," and "to derive them from known chemical and physical causes." One school classified physiology as "organic physics" (Schubert, 1988). Increasingly, the new findings had an important impact on society: Experimental biology was becoming a major social force.*

RESPIRATION

At the beginning of the eighteenth century it had become fully recognized that expired air had lost oxygen, gained carbonic acid and water vapor, and had become hotter. The push was on to measure, as accurately as possible, the quantities involved. To this end, equipment became more and more specialized.

In the 1850s, H. Regnault and J. Reiset made a famous series of respiration studies on animals, to determined oxygen consumption and carbon dioxide production rates, using elaborate but highly functional closed-chamber respirometers (Fig. 17.1). In their technique, carbon dioxide was removed as quickly as it was produced, and oxygen was added as it was consumed, so that the animals were not disturbed by any changes in their ambient air. Regnault and Reiset found that warm-blooded mammals and birds had oxygen consumption rates 10–100 times that of cold-blooded

145

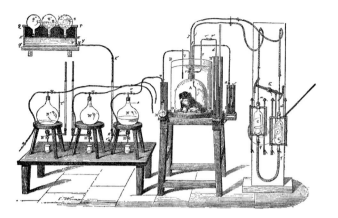

Fig. 17.1. The respiration apparatus of Regnault and Reiset. In this innovative methodology, oxygen was supplied at the rate it was consumed from containers (**N**). Carbon dioxide was absorbed in flasks (**C**).

Fig. 17.2. Dulong's water calorimeter for measuring animal heat production. The calorimeter is double walled for greater insulation; (**D**) contains air.

amphibians and reptiles. They quantified the effect of temperature on oxygen consumption, and noted a scaling relationship between oxygen consumption and size, so that, the larger the animal, the lower the oxygen consumption per kg body weight, and they found that the ratio of carbon dioxide production to oxygen consumption (the respiratory quotient) depended on the type of food consumed. These principles are still taught in basic physiology courses today. A greatly improved method of measuring the amount of heat produced by the living body was provided by Dulong's calorimeter (*see* Fig. 17.2).

Role of Blood in Gas Transport

Lavoisier had concluded that respiration was a form of combustion, in which carbon and hydrogen were oxidized in the lungs to produce carbonic acid and water, while releasing "caloric" (heat). Larange argued that the lungs could not be the site of combustion, because the heat produced would destroy them. He suggested instead that lung oxygen dissolves in the pulmonary blood, and is transported to the general blood pool, where it combines with carbon and hydrogen to form carbonic acid and water vapor. These gases are set free in the lungs, and the temperature increase is buffered by the high heat capacity of water. But Spallanzani

had shown that animals held in pure nitrogen or hydrogen atmosphere, exhaled carbon dioxide, and, from experiments with isolated organs, deduced that the carbon dioxide was not the result of oxygen and carbon combining in the blood, but had come from the tissues.

There were in consequence two rival theories at the turn of the nineteenth century: The combustion theory maintained that a combination of oxygen and carbon took place in the lungs or venous blood; the exchange theory held that oxygen was carried to the tissues, where it was exchanged for carbon dioxide released by the tissues. One difficulty in resolving this issue was that, up until then, there had been no clear demonstration of either free oxygen or carbonic acid in blood.

Some indication that the blood contained oxygen was provided in 1799 by Sir Humphry Davy (1779–1829), who collected a small quantity of oxygen from arterial blood by warming it to 93°C, and noted that this gas could be absorbed by venous blood, with a color change. But the matter was not settled until the 1830s. When using a powerful vacuum pump, Carl Ludwig (1816–1895) extracted gases from whole blood, serum, and isolated blood corpuscles. He found that the oxygen content of serum was similar to that of water, but that whole blood held much more gas.

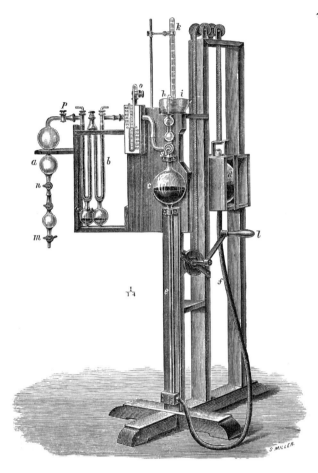

Fig. 17.3. Pflüger's pump for the extraction of gas from blood. (a) Is the blood receiver, (b) are U tubes containing concentrated sulfuric acid to absorb water vapor. The glass globe (c), holding mercury, communicates with similar glass globe (d) via a rubber connecting tube (f). Globe (d) can be raised or lowered. (o) is a pressure gauge.

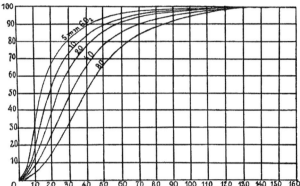

Fig. 17.4. The seminal observation that increasing carbon dioxide reduces the oxygen affinity of blood: The Bohr effect. (From Bohr et al., 1904.)

the blood, with a greater degree of accuracy than before (Fig. 17.3).

Questions arose about how the respiratory gases were held in and transported by the blood. Magnus had concluded that the gases were simply physically dissolved in blood, and that the rate of respiration was determined by the rates of diffusion of oxygen into the blood and carbon dioxide from the blood, which were driven by blood/lung pressure gradients of oxygen and carbon dioxide respectively. But Justus von Leibig (1803–1873) pointed out, in 1851, that the experiments of Regnault and Reiset had shown that animals breathing pure oxygen had the same rate of oxygen consumption as when they breathed ordinary air. The rate of oxygen consumption could not, therefore, be a simple function of oxygen partial pressure (McKendrick, 1889). Investigating the matter, Liebig found that the amount of oxygen dissolved in whole blood was not in simple proportion to the oxygen gas pressure (i.e., did not follow Henry's law). It was concluded that oxygen was not simply dissolved in the blood, but was mainly held in the red blood corpuscles, in a state of loose chemical combination.

In 1857, Julius Meyer (1814–1878) proved that some kind of chemical compound was indeed involved. He found that only the smallest part of the oxygen content is dependent on oxygen gas pressure, and concluded that most binding must be

He in fact found that a fixed volume of arterial blood held 60% volume of gas (all measurements being made at body temperature), and further noted that this gas was one-third (20% volume) oxygen, two-thirds (40% volume) carbon dioxide. These values still stand today. In 1836, using a similar technique, Gustav Magnus found that arterial blood contained more oxygen than venous blood, but that venous blood had more carbon dioxide. Clearly, gas exchange was not occurring in the lungs, but in the tissues, or at least in the capillaries. These findings were supported by E. F. W. Pflüger (1829–1910), using his improved pump for extracting gases from

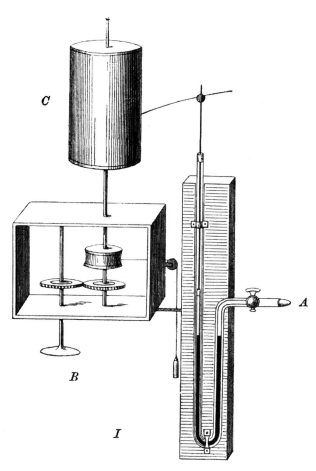

Fig. 17.5. Ludwig's kymnograph. End of manometer (**A**) is attached to an artery, so that the blood pressure changes are transmitted to the marker.

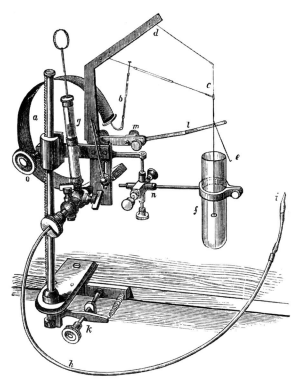

Fig. 17.6. Fick's improved mechanical kymnograph. A hollow spring (**a**) filled with alcohol, connected through levers to rod (**c**). One end of this rod is immersed in oil contained in tube (**f**) in order to dampen oscillations. The syringe (**g**) is used to fill tube (**h**) with saline. (**i**) Is a cannula, and (**l**) is a marker, which can be applied to a moving surface. Various screws are used to fine-tune the position of the apparatus.

chemical (note that the observation was inaccurate, the deduction correct). Meyer also found that blood that had been acidified with vinegar released much less oxygen when heated in a vacuum (observation correct), which he took to indicate that acidified blood bound oxygen more strongly (deduction incorrect). Meyer also found that about 35% of the blood's carbonic acid was held in physical solution and 65% in chemical combination.

The search was on for the blood-oxygen-binding chemical. In 1853, Otto Funke (1828–1879) crystallized the red coloring matter of the red blood corpuscle, called hematin, and found that it contained iron. Left behind was a colorless, iron-free protein, later known as albumen. In 1862, Hoppe-Seyler (1819–1903) noticed an identical spectrum

shift when oxygen was extracted from a dilute solution of hematin, or from diluted blood. He had identified the chemical responsible for the transport of oxygen in the blood, and gave it the name "hemoglobin" (M'Kendrick, 1889) Using spectral analyses, Hoppe-Seyler was then able to demonstrate a loose combination of hemoglobin with oxygen in the red blood cells, where it could exist in two states oxyhemoglobin and deoxyhemoglobin. But the relationship between oxygen partial pressure and blood oxygen content was not established until 1904, when Bohr, Hasselbach, and Krogh published the sigmoidal-shaped blood oxygen dissociation curves, and showed that the oxygen affinity of hemoglobin decreased (i.e., the blood held less oxygen) with increasing carbon dioxide partial pressure or greater blood acidification (Fig. 17.4). The decrease in blood oxygen affinity with acidification is called "the Bohr effect," although there is good evidence that it more justly should be named after the junior author, Krogh (Schmidt-Nielsen, 1991). Advances in the mechanics of blood flow were made through the use of elaborate machines constructed to measure changes in blood pressure in real time (Figs 17.5 and 17.6).

METABOLIC CHEMISTRY

Cell and Protoplasm

At the beginning of the nineteenth century, it was widely believed that, although inorganic compounds (such as salts and acids) could be manufactured by man, inorganic compounds (such as sugars, fats, and nitrogenous compounds such as urea) are so complex that they can only be made by living organisms (in some mysterious way that we cannot now and never will understand). For example, there was a long quarrel between Jean Baptiste Dumas (1800–1884) and Justus von Leibig (1803–1873) over the source of animal fat: Could it be made *de novo* from grain food (Leibig), or does it comes solely preformed from plant material (Dumas)?*

Dumas declared that "plants alone have the privilege of fabricating these products, which animals secure from them either to assimilate them or to destroy them." However, in 1838, Jean-François Peroz, professor of chemistry at the University of Strasbourg, reported that a careful analysis of "pate de fois" geese showed that they gained more fat than the fat in their diet (Holmes, 1974).

The chemical structures of "albumoid" substances (albumin, casein, fibrin—later called "proteins") were a special problem. They were thought to be uncrystallizable, complex chemicals, labile to mild treatment, such as small changes in temperature or acidity, and consequently impossible to structurally analyze. But, in 1828, Friedrich Wöhler (1800–1882) synthesized the first organic compound from inorganic components, preparing urea by reacting lead cyanate with ammonia. He wrote, in an excited letter to Leibig, "I can no longer contain my chemical water, I can make urea without kidney of dog or man." From then on, the highest ambition for investigators into the nature of life was to explain vital activity in physiochemical terms.

Where these "biochemical" processes took place was a matter of intense speculation.

Cell Theory and Protoplasm

In the 1830s, Theodor Schwann (1810–1882), a pupil of Müller, proposed his famous cell theory. While investigating the fine structure of the nervous system, including the thin sheath that surrounds mylenated nerve fibers (the Schwann cell), he came to the conclusion that the small inclusions he observed in these structures were the animal equivalents of plant cell nuclei. On further microscopic investigation, he found that all tissues (muscle, liver, ova, and some blood-corpuscles) contained nuclei. He proposed that these animal nuclei were associated with animal cells, and that these cells are the basic building blocks of all organic tissues, plant and animal. He proudly proclaimed, in his book, *Accordance in the Structure and Growth of Animals and Plants* (1839), that he had "thrown down a grand barrier of separation between the animal and vegetable Kingdoms." Some years later, in 1843, the French microanatomist, Felix Dujardin, obtained a "living jelly" when he mechanically crushed microscopically small animals. This jelly had a "glutinous, structure,

---

*Leibig was one of the few scientists of his time to express concern over prevailing social conditions. Indeed, he has the distinction of being the only physiologist quoted by Karl Marx in *Das Kapital* (1906), in which Liebig is quoted as ironically observing "The laborers in the mines of S. America, whose daily task, perhaps the greatest in the world consists in bringing to the surface on their shoulders, a load of metal weighing from 180–200 pounds, from a depth of 450 feet, live on bread and beans only; they themselves would prefer the bread alone, but their masters, who have found out that the men cannot work so hard on bread, treat them like horses, and compel them to eat beans; beans, however, are relatively much richer in phosphate of lime than bread."

insoluble in water," and contracted into "spherical masses." He called it "sarcode," and thought that it might be the essential living part of the cell. The idea became quickly and widely accepted that the animal cell consisted of a central dot, the nucleus, surrounded by a structureless, gelatinous material, the sarcode (later called "protoplasm"), bounded by a membrane (Magner, 1979).

Initially, cell formation was thought to occur through a process analogous to crystal growth, in which the cytoblastema formed around the nucleus, out of the bathing solution, or mother liquor. But the crystallization theory was overthrown in the mid eighteenth century, when Rudolph Virchow (1821–1902) demonstrated that every cell is the product of a preexisting cell.

Schleiden and Schwann defined the cell as a "smooth vesicle with a firm membrane enclosing the fluid content", and Schleiden attached particular importance to the nucleus. But it was argued that the attributes of life rest in the protoplasm, since not all animal cells could be seen to have membranes or a nucleus. But better microscopes later resolved this objection.

The protoplasmic theory was gradually developed during the 1860s, when all of the vital processes, such as respiration, assimilation, growth, muscular contraction, and nerve conduction, were attributed to the single protoplasmic entity. In 1868, T. H. Huxley (1825–1895) declared that "all vital actions may be said to be the result of the molecular forces of the protoplasm which display it" (Eldon, 1965). Protoplasm was thought to be a single homogeneous substance, or a giant molecule with numerous side chains, each with a special metabolic function, and speculations abounded concerning the chemical mechanisms involved. Pflüger, for example, thought that the special properties of protoplasm resulted from it having certain "unstable" groups, such as aldehydes. In 1876, he suggested that protoplasm contains an energy-rich protein with CN, which combines explosively with oxygen to liberate carbon dioxide. These explosions cause intramolecular vibrations, which produce the cellular biological effects.

Enzymes and Catalysts

In the early eighteenth century, the idea appeared that particular chemical changes in the body might be caused by special biological chemicals. It was soon established that an acidic fluid was secreted when meat was placed in the stomach, but acid by itself was insufficient, because it had no effect on the digestion of "albuminoids". In 1834, Johan N. Eberle (1787–1837) noted that the stomach contents of rabbits was covered with mucus during digestion. He extracted the mucus, mixed it with fibrin, albumen, and casein, and "with an indescribable joy," as he noted in a letter to Müller, he saw that they were "completely chymified." Müller and Schwann haled this finding as "a brilliant discovery" (Holmes, 1974). Eberle had shown that gastric digestion was brought about by stomach acid and something else, some factor in the gastric mucus (mucus from other tissues were tested and found to have no digestive capacities). The gastric mucus factor was present in very small quantities, and was not proportional to the product, and, most significantly, it did not appear to be consumed in the reaction. Schwann confirmed and extended Eberle's findings and gave it the active compound name "pepsin." The first "biological catalyst" had been discovered.

Were ferments (later called "enzymes") also responsible for chemical changes inside the cell? Ferments were characterized as possessing "the power even when present in small quantities of producing great changes in the bodies without entering themselves into the change" (Huxley, 1881).

By the middle of nineteenth century, there was some suspicion that there might be some common chemical mechanism behind the production of lactate in contracting muscle and the appearance of acid in the fermentation of wine. A controversy arose over whether alcohol fermentation was a vital (dependent on some life force) or a chemical process. In 1839, Leibig came down firmly on the side of nonliving processes, and proposed that fermentation was caused by albuminoid matter being decomposed by oxygen, and that the splitting caused molecular vibrations that mechanically

Fig. 17.7. Sketch for a physiological experiment, from Claude Bernard's note book. (From Hoff et al., 1976.)

cleaved sugar to alcohol and carbon dioxide (Kohler, 1972). However, Louis Pasteur (1822–1895) showed that fermentation did not need oxygen, indeed that it proceeded better without it. In 1875, Pasteur announced his germ theory of fermentation, and established that the yeast cell is responsible for the conversion of sugar to alcohol in wine making. He tried to extract the cell juices responsible, but failed, and concluded that fermentation was a vital process, requiring the intact cell.

Pasteur's prestige and authority was so great that the matter of "vital" fermentation appeared settled. And so it rested for the next 30 years, until 1897, when Eduard Buchner (1860–1917) submitted yeast to great pressure, and succeeded in initiating fermentation with a yeast cell extract, proving that fermentation did not in fact require the intact cell. He called the responsible agent that he isolated, a substance dissolved in the cytoplasm, "zymase" (zymase turned out to be a mixture of 12 separate catalytic processes).

Buchner was awarded the 1907 Nobel Prize in chemistry for his discovery. But reexamination showed that Pasteur's findings were not in error. It appeared that, by bad luck, the Parisian strain of yeast that Pasteur had worked with lacked sucrase, the enzyme that initiates the pathway of sugar metabolism. The Munich yeast contained ample amounts of this enzyme.

In the late nineteenth century, Emil Fisher (1852–1919) proposed an imaginative lock-and-key theory of enzyme action, and, beginning about 1879, the protoplasm theory began to be replaced by the theory that chemical changes in the cell are brought about by many highly specific intracellular enzymes. Molecular biology was born (Fruton, 1996).

INTERNAL SECRETIONS

Claude Bernard (1813–1878) deserves special attention as one of the most innovative and imaginative physiologists of the nineteenth century. He was a prolific writer. His book, *Introduction à l'étude de la médecine expérimentale* (1865), became a landmark for physiologists, and his *Leçons sur les phénomaenes de la vie communs aux animaux et aux végétaux* (1878) was extremely influential.

Bernard was a staunch antivitalist, and argued that only one chemistry accounted for the living and nonliving. He strongly believed that all vital phenomena had an ultimate physiochemical basis, which could only be revealed through animal experimentation (Fig. 17.7). He distinguished between the "external environment," which the animal experiences and cannot control, and an "internal environment," where the tissues and cells live. Developing this idea, he arrived at the seminal concept of the constant "internal milieu." "Life does not run its course within the external environment … but within the fluid internal environment, formed by circulating organic liquid that surrounds and bathes all the anatomical elements of the tissues." The external environment might change, but the "constancy of this internal environment is the condition for free and independent life." In order to maintain its role of supporting life, the composition of this "milieu interieur" (the blood and the tissue-bathing extracellular fluid) must be constrained by continuous compensatory regulation. The task of physiology was to characterize the milieu interieur and identify the mechanisms responsible for its maintenance.

In 1932, Walter Cannon (1871–1945), who was important in promoting Bernard's ideas to "Anglo-science," coined the universally accepted term "homeostasis," to describe this "relatively constant" internal condition.

Bernard made fundamental discoveries through bold experimentation, using mostly rabbits and dogs, but also other species. His approach to the use of experimental subjects was pragmatic. "There are also experiments in which it is proper to choose certain animals which offer favorable anatomical arrangements or special susceptibility to certain influences. This is so important that the solution of a physiological or pathological problem often depends solely on the appropriate choice of the animal for the experiment so as to make the results clear and searching".

Bernard's studies on the role of the liver in carbohydrate metabolism are brilliant. As we have mentioned, at that time, it was widely thought that plants and animals had separate but complementary metabolisms. Plants synthesized complex organics from simple inorganics; animals broke them down by feeding on the plant organics, and incorporated them into their tissues (Larner, 1967).

Bernard was especially interested in where blood glucose (at that time, thought to be solely a plant-made material) came from. He put a fasting dog on a meat-only diet, and, to his surprise, found that the hepatic vein of the liver was rich in glucose. Further experiments of the same kind, on dogs that had the liver portal, mesenteric, and pancreatic veins ligatured, showed that glucose only appeared in the liver portal vein after the fasting dog had been fed meat. Perhaps, he queried, the dogs were making glucose in the liver, in the absence of a source of plant-derived glucose?

Pursuing these findings with his rigorous experimental approach, Bernard investigated whether the liver was actually producing glucose. He analyzed

a freshly removed liver for its glucose content, then, to check his measurements, he reanalyzed it the following day. To his surprise, the glucose content had increased: It appeared that sugar was forming in liver, even when it had been removed from the body. He flushed the liver with water through the portal vein, until no more glucose appeared in the washout, then repeated his analyses several hours later. Glucose reappeared. Bernard concluded that glucose is not formed in the blood of the liver, but comes from a precursor in the tissue that was not washed out by the flushing.

He next tackled the nature of this liver glucose-forming insoluble substance. He flushed the liver with hot water and added alcohol to the extract: A white precipitate was produced, which, under the microscope, was made of grains that closely resembled plant starch, which was confirmed by chemical analysis. Bernard had discovered animal starch, or glycogen. This organic glycogen was therefore synthesized in the liver, and glycogen was broken down to glucose in the liver.

These experiments also demonstrated that the liver had at least two secretory functions: the long-known one of secreting bile, and Bernard's newly discovered role of secreting glucose. This concept challenged the cherished old dogma of one organ, one function. Furthermore, Bernard had discovered a new class of "internal" secretions, initiating a path that was to lead to the discovery of hormones.

RENAL PHYSIOLOGY

Before the structure–function relationships of organs were understood it was thought not unlikely or impossible that their secretions could sometimes issue from other routes. For example, although the kidney was well recognized as the prime source of urine, in the absence of any understanding of renal mechanisms, there was no compelling reason to believe that the kidney was the sole source. This commonly accepted concept is illustrated in a communication by Dr. Ambrose Dawson to the Royal Society in 1759, which relates the case of a young lady who compensated for "a total suppression of

16 Plan ___ Proportions as in Man.

Fig. 17.8. Bowman's illustration of the human nephron. The Bowman's capsule (**c**) holds a tuft of capillaries.

urine for above one year" by discharging "urinous like" fluid from her nipples and by vomiting urine. It was similarly thought that, under certain conditions, such as overabundance or stoppage of the natural outlet, semen and menstrual blood could also be expressed through the skin or other pathways.

For a long time after Malphigi's description of the microanatomy of the kidney in the seventeenth century, not much more was known about kidney fine structure, or about how the tubules and blood vessels related. The problem was that the delicate tissue easily became mush when it was squeezed for microscopic observation. William Bowman (1816–1892) had an ingenious solution. He injected a solution of potassium bichromate into the vascular system, followed by a solution of lead acetate. This resulted in the precipitation of a fine

powder of lead dichromate throughout the kidney vascular system, allowing small blood vessels to be outlined with clarity (Fine, 1987).

Bowman then embarked on a painstaking examination of the microstructure of kidney tissue of a wide variety of vertebrates, including fish, frogs, parrots, lions, and humans. He made thin microscopic slices, and the tissues were carefully teased apart under the microscope. In 1842, he published the classic *On the Structure and Use of the Malphigian Bodies of the Kidney, with Observations on the Circulation Through that Gland*, for which he received the Medal of the Royal Society. Bowman's major finding was the discovery of a cap of blood vessels, the Bowman's capsule, which engulfs, but does not enter, the termination of the kidney tubule (Fig. 17.8). He hypothesized that some of the substances in urine were the result of secretion from the blood into the tubules.

Ludwig (1843) quickly and strongly disagreed with this concept of secretion, declaring it too fanciful, too "vitalistic." Instead, he proposed a mechanistic filtration model of urine production, powered by cardiac-derived hydrostatic pressure in the glomerular capillaries. Ludwig hypothesized that the glomerular capillaries are permeable to all substances in the blood, except proteins, lipids, and cells, and, under hydrostatic pressure, a protein-free and cell-free filtrate is produced in the terminal kidney tubule. However, since many substances, such as urea, are found in much higher quantities in the urine, compared to the blood, a substantial portion of the filtrate must be reabsorbed during its passage through the renal tubules.

But how the kidney could produce a urine more concentrated than the blood remained a mystery. Jacob Henle (1809–1885) discovered the loop arrangement of the medullary tubules, which eventually proved to be the structural grounds for an explanation of the concentration of urine through a counter-current system.

In the 1850s, the filtration hypothesis received wide support from the use of a variety of dyes, which confirmed that the glomerular capillaries were permeable to small molecular substances. But the argument continued. In 1874, Rudolph Heidenhain (1834–1897) observed the dye, indigo carmine, in the cells and luminal renal tubules of rabbits whose urine formation had been arrested by ligature, but no dye color was discernable in the Bowman's capsule. He concluded that the dye had to have been secreted into the urine, independent of glomerular filtration. The Bowman-Heidenhain theory of secretion rivaled Ludwig's filtration theory for many years. It was not until the 1920s that glomerular micropuncture studies, pioneered by Alfred Richards, confirmed filtration, and in 1927, Marshall unambiguously demonstrated secretion in the glomerular-lacking kidney of the monkfish, Lophius. Both sides were right.

DIGESTION

In the early eighteenth century, it was thought that gastric juice caused food in the stomach to be reduced to a pulpy fluid called "chyme," through a process called "chymification." Contractile motions of the stomach sent the chyme to the small intestine, where it "mingled" with juices, "such as bile from the liver and pancreatic juice from the pancreas." This caused the "truly nutritious parts" of the chyme to separate and form a milky liquid called 'chyle," through the process of chylification. Chyle was believed to be composed of minute, colorless globules, which were taken up and carried into the blood by means of the lymphatic vessels, "whence taken up and mingled with the blood to become part of the living body" (Agassiz and Gould, 1851). Lymph globules were identified as one of the microscopial components of the blood.

Not much was known about the active agents in digestion. Schwann had discovered pepsin in gastric juice, and bile was thought to be a "natural digestive." Bile had traditionally been seen as an excretory product. But the idea that it might have a special role in digestion took hold in the eighteenth century For example, in 1732, John Arbuthnot noted that ordinary folk used ox bile in washing clothes, and that it was particularly effective for removing grease marks. Perhaps, thought Arbuth-

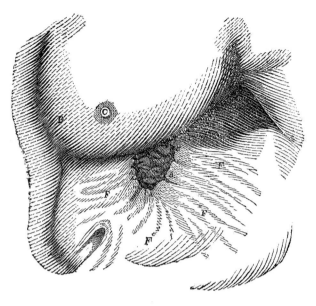

Fig. 17.9. Abdominal hole opening directly into the stomach of Alexis St. Martin, which allowed Dr. Beaumont to perform his seminal experiments on gastric digestion, to the discomfort of Mr. St. Martin. (From Beaumont, 1833.)

not, bile is "a saponaceous substance and ... mixeth the oily and watery parts of the aliment together".

But experimental investigations had to wait until the nineteenth century. Some of the most interesting and important were conducted by William Beaumont (1785–1853), a doctor in the United States Army. In 1822, a fur trapper, Alexis St. Martin, who had been accidentally shot in the stomach, became Beaumont's patient. Although the patient recovered, a permanent opening was left into the man's stomach (Fig. 17.9). In 1825, Beaumont commenced an eight-year series of experiments, detailed in his book, *Experiments and observations on the gastric juice and the physiology of digestion* (1833). In his first experiment, he reports that, at 12 o'clock, "I introduced, through the perforation, into the stomach, the following articles of diet, suspended by a silk string ... so as to pass without pain -viz. A piece of high seasoned a la mode beef, ... a bunch of raw sliced cabbage... At one o'clock withdrew and examined ... found cabbage about half digested, meat unchanged. Returned them into the stomach."

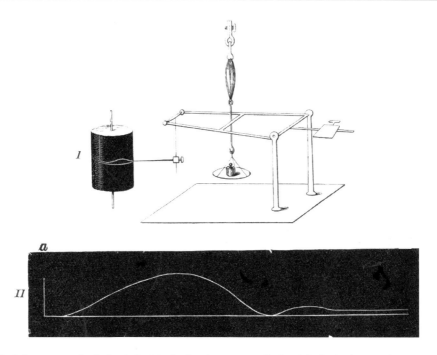

Fig. 17.10. Helmholz's myograph. A short electrical stimulus was applied at (**a**). A single muscle contraction results; the second wave is an artifact.

Hundreds of experiments followed on the long-suffering St. Martin, putting different foods into the abdominal hole and withdrawing them after certain periods of time. Beaumont mixed olive oil or cream with gastric juice, and noted that the fat was more dispersed when the juice was "strongly bile-stained" than when it was clear. He inferred that, "when oily food has been used, bile assists its digestion."

However, the crucial direct evidence on the role of bile in lipid digestion were given by Schwann. He made a fistula from the bile duct to the skin of a dog, which caused the entire biliary output to be diverted from the intestine. Schwann noted that, although ravenously hungry, the dog would lose weight when deprived of bile, and that the feces it produced were bulky, white, and rich in fat.

The Nervous System

NEURONAL CONDUCTION

In the early nineteenth century, interest in animal electricity ebbed, and no important advances were made. This situation changed when the Ital-
ian, Carlo Matteucci (1811–1863), built a special galvanometer for measuring current on muscle surface. In his investigations on muscles of cold- and warm-blooded animals he found that this biological current could transfer from one muscle to another through an attached nerve, and caused contraction in the second muscle. He also noted that the muscle current decreased during tetany.

In 1811, Charles Bell (1774–1842) privately published his discovery that the sensory fibers enter the spinal cord dorsally, and that the motor fibers leave the spinal cord ventrally. The French physiologist, François Magendie (1783–1855), independently made the same discovery. The fact that nerve conduction was seen to normally travel in one direction became a fundamental concept for theories of nerve function.

Johannes Müller (1801–1858) thought that the propagation of the nerve impulse was so fast that it was comparable to the speed of light. This belief was supported by introspection, which does not show any lag in the perception of motion of one's foot and the conscious act to move it (Boring,

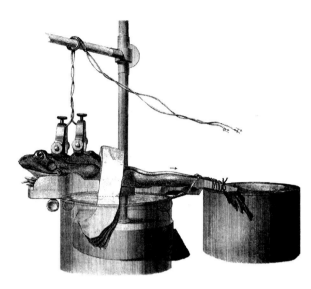

Fig. 17.11. Du Bois' demonstration of electrical stimulus applied to the nerve transmitted to muscle (1848). The frog is tied, with nonconducting silk, on a wooden platform. Electrical current is applied to the spinal cord by means of two brass clamps.

1957). In 1840, Müller wrote, in his *Handbuch der Physiologie des Menschen*: "We shall probably never attain the power of measuring the velocity of nervous action; for we have not the opportunity of comparing its propagation through immense space, as we have in the case of light."

Only 10 years later, using the newly invented galvanometer, his pupil, von Helmholtz (1821–1894), estimated the speed of nervous conduction in a frog to be about 30 meters per second (Fig. 17.10). But there was a new problem. The great difference in speed between ordinary electric conduction through a wire and stimulatory propagation along nerve fibers presented a conceptual difficulty. The speed of conduction in nerve seemed far too slow for electrical propagation.

Inspired by Matteucci's findings, Emil du Bois-Reymond (1818–1896), assistant to Müller, built a more sensitive galvanometer, and made the key observation that both muscles and nerves produce electrical currents during activity (Fig. 17.11). He devoted his energies to investigating the phenomena, and, in 1848, published the first of two volumes devoted to animal electricity: *Untersuchungen über tierische Elektrizität*. Du Bois-Reymond believed that Matteucci's "muscle current" was the result of a potential difference between the negative inner and positive surfaces of muscle at rest, and he demonstrated that the excitation stimulus passing along the nerve is not a current flow, but a wave of changes in potential (*Schwankung*). Du Bois-Reymond believed that he had identified the principle involved, and surmised, in 1872, that the electrical potential was caused by electromotive molecules in the nerve, arranged in series, end-to-end, in an unbroken circuit. He called this stimulatory potential the "negative Schwankung," known today as the "action potential."

In 1868, Julius Bernstein (1838–1911), a student of du Bois-Reymond, built an analyzer that produced the first plot of the action potential, and launched a long series of electrophysiological measurements under different conditions. He found that the nerve potential was dependent on the chemical composition of the bathing fluid, especially the potassium concentration. He proposed, in 1902, that the resting electromotive force was based on the selective permeability of the nerve membrane to potassium ions, and was a direct function of the transmembrane potassium gradient. He theorized that nerve conduction is a wave of electrical depolarization caused by changes in membrane resistances. Ernest Overton (1865–1933) had found that calcium and sodium were necessary for nerve excitability, and suggested that some sort of sodium-potassium exchange might be involved in signal conductance. But it was recognized that, to make any further substantial advances in this field, it would be necessary to measure the transmembrane potential directly. The breakthrough came in 1936, when J. Z. Young described the giant squid axon, which was so large (1 mm diameter: 50–1000 times that of the mammal) that it could be impaled directly by microelectrodes. The absolute values of the resting potential could then be measured, as well as the dynamic changes of the action potential. Electrophysiology became a rich field of discovery.

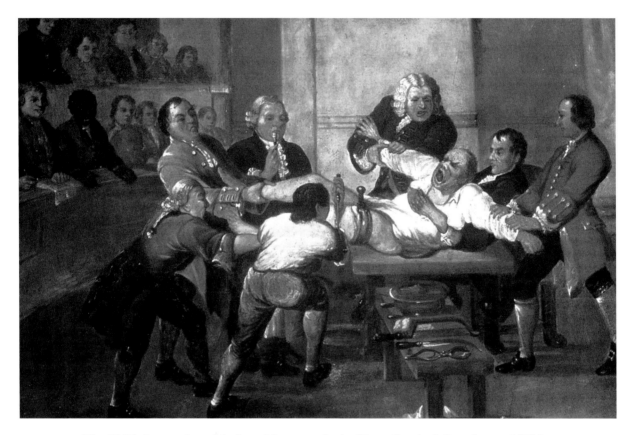

Fig. 17.12. Amputation of the leg without anesthetic. (From Gersdorf, Strassbourg, 1517.)

CEREBRAL LOCALIZATION

The new techniques for measuring bioelectrical phenomena led to the final discarding of the concept of the brain as a secretory organ, and to it being viewed as an electrical organ. Since Greek times, the brain was believed to be the site or home of mental functions, but the long-running question was about the sites of the different faculties. From medieval times until the seventeenth century, the hollow ventricles were considered the houses of different brain functions, the brain tissue merely serving as the passive walls of the container. But, in the seventeenth century, there was a gradual realization that the working material of the brain was the brain matter itself. For example, in 1664, Willis suggested that the cerebrum was responsible for voluntary actions and the cerebellum controlled involuntary movements. Haller undertook a long series of experiments on localized brain

function, irritating different parts of the brain to see if different muscle masses were stimulated, but the results were inconclusive and unsatisfactory.

Nevertheless, the idea that different mental functions were highly localized in the brain was widely accepted. It was also thought plausible that the degree of development of a particular part of the brain determined the power of the locally sited mental faculty. This idea was refined into the science of phrenology, which claimed that details of an individual's mental character could be revealed through an examination of the surface contours, the bumps, on the skull.

The champions of phrenology, the neuroanatomist Franz Joseph Gall (1785–1828), and his pupil J. C. Spurzheim (1776–1832), believed that the brain surface takes its shape according to the degree of localized function it exercises (Marshal and Magoun, 1998). From this premise, Gall argued

Fig. 17.13. Administering chloroform. (From Snow, 1858.)

that the brain's topography determined a person's "affective and intellectual activities," his behavior, thought, and emotion, and provided detailed examples of these connections. For example, he claimed, in his book, *On the Functions of the Cerebellum* published in English in 1838, that he had collected a "prodigious number of facts," which suggested that "a connection might exist between the functions of physiological love and those parts of the brain situated at the base of the neck". The cerebellum, he declared, was the source of the generative instinct, or "amativeness."

Spurzheim recognized 37 powers of the mind, which corresponded to an equal number of "organs" of the brain. Although scoffed at by most professionals, this simple theory became popular, and a phrenology industry swept Europe and America in the nineteenth century, surviving into the early twentieth century. At one time, Britain had 27 phrenological societies (Boring, 1957). Professional journals were established, such as the *American Phrenological Journal*.

However, in the latter half of the nineteenth century, real advances in brain function localization were being made, from careful examination of the consequences of clinical damage in humans and experiments in animals. Localized function was demonstrated in 1870 by Gustav Fritsch and Edward Hitzig, who showed that local electrical stimulation of the exposed cortex of dogs induced movement in contralateral limbs. And from pre- and postmortem observations on a series of patients, Paul Broca (1824–1888) localized a center for the ability to articulate language (motor aphasia) in a precise region of the brain, the posterior part of the third frontal convolution of the left hemisphere, now known as "Broca's area." In 1874, Carl Wernicke (1848–1904) identified the locus of cerebral injury responsible for sensory aphasia, in a discrete locus in the sylvian fissure of the left hemisphere, or "Wernicke's area."

BENEFITS

For the first time, the advances in biology started to have substantial benefits for human welfare: The most dramatic concerned antiseptics and anesthetics.

Antiseptics

Until the nineteenth century, many diseases were believed to be transmitted by a noxious vapor, the miasma, which emanated from putrescent materials and swamps, and was carried in the air. This bad air, malaria, had to be avoided for any hope of recovery.

Louis Pasteur found that the souring of milk, wine, and beer was caused by heat-sensitive microorganisms, and that these "germs" could be killed by heat (pasteurization). In 1865, he discovered that microorganisms were responsible for the silkworm disease, pébrine, which was devastating the French silk industry. Many scientists were attracted to the

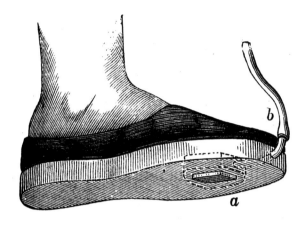

Fig. 17.14. Marey's apparatus for recording head and foot movements in a running man. (From Marey, 1874.)

germ theory of disease, i.e., that specific infectious diseases were caused and transmitted by specific microorganisms. Great advances were made as the nineteenth century drew to a close. Robert Koch

(1843–1910), for example, made a pure culture of anthrax bacillus in 1867, and isolated bacillus responsible for tuberculosis in 1882. The germ theory of disease produced new strategies of defense. The surgeon Joseph Lister (1827–1912), was particularly interested in healing. At that time, mortality following amputations often exceeded 50%, and gangrene was rampant. Lister argued that wounds do not need to secrete pus to heal. Inspired by Pasteur, he advocated cleanliness in the operating room, and settled on and championed carbolic acid spray as a killer of germs. The carbolic acid spray was first used in 1865, and published in *The Lancet* in 1867. The use of antiseptics was rapidly and widely adopted. By the end of the nineteenth century, the isolation of disease-causing microorganisms had stimulated the new science of immunology and the discovery of specific antibiotic chemicals.

Anesthetics

The pain of surgery was excruciating (Fig. 17.12), and the success of an operation depended on the speed and strength of the surgeon. In the early nineteenth century, Langenbeck, surgeon-general to the Hanoverian army, could reputedly amputate a shoulder as quickly as one could "take a pinch of snuff" (Haggard, 1929). The new chemistry of the nineteenth century produced new synthetic chemicals, and tests on their biological effects followed almost immediately. One of the first examples of chemical anesthesia was provided by Davy's nitrous oxide. "Laughing gas" not only provided social fun, but was found to deaden pain, and Davy suggested that nitrous oxide "may be used with advantage in surgical operations." In 1844, an American dentist, Horace Wells, rather heroically, but painlessly, pulled out one of his own teeth, while under the influence of nitrous oxide.

The new organic chemistry produced new volatiles that had anesthetic effects. One of the first, ether, caused unconsciousness, but was distasteful and dangerous, because it was easy to overdose. In 1837, Leibig and Wöhler discovered chloroform.

Fig. 17.15. Beginnings of high altitude physiology. The hydrogen balloon allowed a rapid ascent to high altitudes. On 5 September 1862, James Glaisher (at right), chief meteorologist at the Royal Observatory at Greenwich, recorded the onset of physical and mental impairment brought about by the increasingly severe hypoxia, until he collapsed unconscious at an altitude of seven miles. Some of the accompanying pigeons appear unperturbed.

Just 10 years later, it was first used in childbirth by James Young Simpson (1811–1870), professor of midwifery at Edinburgh University.

Chloroform quickly became the anesthetic of choice because it was safer than ether, and methods were invented for its controlled application (Fig. 17.13). But there was an immediate storm of protest: Anesthetics might be acceptable for the pain of toothache, and so on, but not for childbirth. Simpson was roundly condemned by the Scottish clergy in sermon and in print (Haggard 1928). "Chloroform is a decoy of Satan ... contrary to the principles of religion." It flew in the face of the curse of Eve in *Genesis* III:16: "Unto the woman ... in sorrow shalt thou bring forth children." Simpson was also attacked professionally by a famous (at that time) Dr. Meigs of Philadelphia, who argued that labor pain was "natural" or "physiological," and should not be interfered with, and is therefore "a desirable ... manifestation of life force." The unrelenting attacks were so strong that they required a response. Simpson's scathing reply, *Answers to the Religious Objection Against the Employment of Anesthetic Agents in Midwifery and Surgery*, was published in 1847, in which he sarcastically asked such questions as, why was the pain of labor "natural," but toothache not. However, chloroform-assisted childbirth was not respectable until Queen Victoria used it at the birth of Prince Leopold on April 7, 1853.

NEW FIELDS

As the century advanced, new technologies opened up new fields for research, such as the mechanics of locomotion (Fig. 17.14) or the new and dangerous territory of high-altitude physiology (Fig. 17.15).

18 Evolution and Physiology

Although Earth had been displaced from the center of the material universe by Copernicus in the fifteenth century, man had remained at the apex of the biological realm. Until the beginning of the nineteenth century, many biological scientists, including most physiologists, believed that they were not just trying to explain how nature worked, but that they had a higher mission: to reveal God's design in nature. This position radically changed as the century advanced.

An important shift in scientific perspective came from the realization that biological time was immensely greater than had been previously thought, and that the assemblages of life forms in the past were very different from those of the present. The idea that the present was descended from the past, contrasted to seeing it as a series of *de novo* creations, became more plausible. Darwin's theory of evolution was enormously influential, not only in science, but also in society.

Around the beginning of the nineteenth century, science was still viewed by many as having a deeply spiritual element: to answer the larger questions connecting man to the absolute, and to illuminate the Christian God's creation of the universe and the laws that governed it (James, 1993). But the hard belief, based on many scholarly interpretations of the Bible, that Earth was about 6000 years old, was being challenged by the new geologists. The most important, James Hutton (1726–1797), showed that the fossil record revealed that Earth had been inhabited by a succession of different animals and plants;

Charles Lyell (1797–1875) convincingly calculated that Earth's rock strata had taken many thousands of millions of years to form. The biblical time-scale was gradually abandoned by even conservative scientists. For example, although still regarding each species as an individual creation, the famous French naturalist George Buffon (1707–1788) maintained that life had been on earth for hundreds of thousands of years, and that the biblical seven days of creation represented seven epochs of indeterminate length, man appearing in the seventh (Magner, 1979).

Louis Agassiz (1807–1873), one of the most prestigious biologists of his time and an expert on the description and classification of fossils, believed that each geological epoch, such as the age of the fishes, the age of the reptiles, or the age of the mammals, represented an "immense period of time." In his book *Outlines of Comparative Physiology* (1851), he declared that there is "manifest progress" in the succession of beings on earth, in that they become closer and closer to the living fauna, and ultimately to humans.

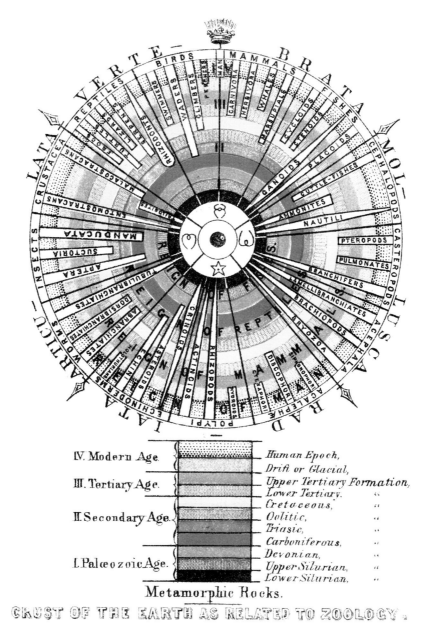

CRUST OF THE EARTH AS RELATED TO ZOOLOGY.

Fig. 18.1. Louis Agassiz's "physiological" interpretation of God's plan on the unfolding of life on earth for the preparation of man. (From Agassiz and Gould, 1851.) (*See* color plate appearing in the insert following p. 82.)

But, he admonishes, there is nothing like parental descent connecting them, and close inspection reveals a divine plan (Fig. 18.1). The Paleozoic fishes are in no respect the ancestors of the reptiles: "the link by which they are connected is of a higher and immaterial kind.... Man is the end towards which all animal creation has tended," and "to study the succession of animals in time ... is to become acquainted with the ideas of God himself."

But what was this plan or design in the seemingly bewildering array of animals and plants? In the eighteenth century, Carl Linneaus (1762–1788) produced a system of classification that became widely accepted. Animals and plants were arranged in a tree-like pattern, according to the degree of shared morphological features. A little later, William Sharp MacLeay (1792–1865) produced a competing system (quinarianism) in which, using analogies, animals were arranged in groups of five circles, circles being the most perfect form and closest to God. The leading advocate of the quinary system, William Swainson (1789–1855), accommodated the most recent discoveries of fossil reptiles in this "natural Divine pattern," which could even predict gaps, or missing "links of a chain," to be filled by future discoveries (Swainson, 1838; Fig. 18.2).

In 1859, Darwin published his bombshell, the *Origin of the Species*, arguing that current life forms evolved from common ancestors, and that the mechanism for change was natural selection. The displacement of humans from the center of animal creation was his most subversive conclusion. In Darwin's theory, humans become just another species, one that has evolved from an anthropoid ape stock. This idea was received as outrageous blasphemy by the establishment, and by many conservative scientists. For example, Sir Richard Owen (1804–1892), curator of the London's Hunterian Museum and scientific founder of the Natural History Museum, was an impassioned opponent of natural selection and particularly rejected the demotion of humans' top animal status. As proof, he argued that there were several anatomical features unique to the human brain, the most important being the "hippocampus minor," and that it was this structure that provided the unique human intellectual attributes, not "evolutionary' history." Owen claimed that it was even absent, or virtually so, in an idiot human (Gross, 1999). A well-publicized, acrimonious debate ensued, in both the scientific and popular press, which was only resolved when Thomas Huxley demonstrated that the so-called "hippocampus minor" also existed in monkeys and apes (Fishman, 1997).

The idea that the human species has come about through blind chance was deeply disturbing. Even

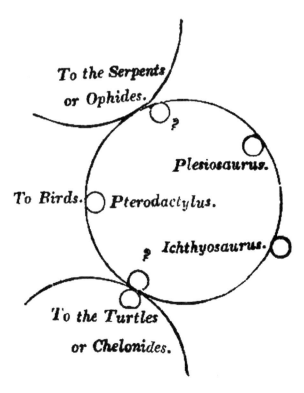

Fig. 18.2. The Quinarian system of natural classification, based on circular analogies. (From Swainson, 1838.)

Alfred Wallace (1823–1913), the co-discoverer of evolution through natural selection, was uncomfortable with God being taken out of the evolution of the universe. He believed that a Designer must be involved in the marvelous order of nature. He argued in his book, *The World of Life* (1914), that there are some biological phenomena that cannot be accounted for by natural selection. "It is almost inconceivable that the exquisite fragrance developed only by roasting the seed of the coffee shrub should be a chance result of the nature of the juices essential for the well being of this particular species." Moreover, he argued, recent discoveries in the biological sciences have revealed complex biological structures that do not yield to materialistic explanation. These included protoplasm, which "is so complex a substance … that it is quite beyond the reach of chemical analysis," and "the atomic structure of … haemogobin is almost impossible of determination" (Wallace, 1914).

Ironically, hemoglobin was the first enzyme-like protein whose structure and related function was fully described. Arguments in an identical vein are still made by present-day creationists. For example, the creationist biochemist, Michael J. Behe maintains that the processes of blood clotting are so intricate that they could only have come about by design, not evolution (Behe, 1996), and holds that it is inconceivable that the complexity of molecular biology can "admit of a natural explanation": "only design can explain ..." and so on.

Yet, in one aspect, Wallace's appreciation of nature was far ahead of his time. He argued that we have a duty to posterity to preserve virgin forests, and that "to pollute a spring ... or to exterminate a bird or beast, should be treated as moral offences and as social crimes" (Wallace, 1914).

But, in spite of deeply felt opposition from a comparative few, Darwinian evolution found quick general acceptance, and today, evolution is the great unifying theory of modern biology, from ecology to molecular genetics. Physiology seeks to explain how the body works, and how it copes with environmental changes. The new concept of functional adaptations in physiology, evolving through natural selection, promises to provide quantitative explanations of biological design (Diamond, 1993).

In Britain, in particular, an evolutionary perspective was quickly applied to physiology. The Cambridge School of Physiology, under Michael Foster (1836–1907), was particularly influenced by evolutionary concepts (Geison, 1978). In his book, *A Textbook of Physiology* (1877), he refers to the "dominant principle" of a physiological "division of labor," for the body to function as a whole, a phrase that he probably borrowed from Darwin (Geison, 1987).

Walter Gaskell (1824–1878), for example, studied "the evolution of function" in order to understand the complexities of rhythmic motion. He thought that he could unravel the complicated system of myogenic and neurogenic control of the human heart through investigations on simpler models, such as frog and turtle hearts. This "evolutionary" approach encouraged a broad comparative outlook in British physiology, and an interest in the interaction of the organism with its environment. Comparative physiologists studied such problems as how desert animals conserve water, or how some diving vertebrates can hold breath for hours. Huxley discussed the origin, types, and function of blood corpuscles in vertebrates (Huxley, 1872).

The finding that the relative mineral composition of the blood of most animals resembles that of sea water encouraged speculation that this feature of the internal milieu reflected a marine origin of the common ancestor that had been modified through time, as species diverged, but had not changed in its essence (Macallum, 1926).

It has been argued by Geison (1978) that the fresh perspective provided by the evolutionary approach contributed to the rejuvenation of late Victorian physiology in Britain, which had been overshadowed through much of the nineteenth century by France and German science. In the late nineteenth and early twentieth century, British physiologists again were at the forefront, making major contributions to such new fields as endocrinology and chemical coordination and integrative neurophysiology. And the evolutionary approach has remained fruitful: Evolutionary physiology is behind much of current quantitative genetic analysis and metabolic control theory (Garland, 1985).

19 Physiology Abused

CONTENTS

By the beginning of the twentieth century, biology was a firmly established mature science. The subsequent exponential growth in understanding how biological systems work produced revolutionary improvements in medicine and agriculture. But increased power to do good carries the downside of a greater ability to harm. This chapter focuses on the misuse of biological science in the twentieth century, with examples from the trivial to the tragic.

Viewed from the perspective of the twenty-first century, the seminal discoveries at the beginning of the twentieth century were: In physics, the discovery of radioactivity (through radiation), which quickly led to finding subatomic particles and the recognition of the equivalency of matter and energy; in biology, the discovery of the cellular units of inheritance, and, more slowly, the elucidation of their chemical basis in molecular biology. By that time, the principles of scientific investigation had been fully established, an endpoint for this book. As the twentieth century advanced, experimental biology continued to grow at an exponential rate, and disciplines break up into further and further subspecialties. University schools devoted to the new biological sciences were established and specialized scientific journals flourished. Indeed, since 1966, the web-

site, Medline (http://biomednet.com/db/medline), lists the publication of more than one half million papers on "heart," and almost 300,000 on the highly specialized topic "sodium channel." Trying to cover in a single chapter the advances in biochemistry and physiology in the twentieth century is pointless; many volumes would be required even for a decent overview. This chapter deals, instead, with an increasingly important theme that has been an undercurrent throughout the book: The influence of biological theory on society. It will touch briefly on the gullibility of some eminent scientists, and the misplaced overconfidence in psychosurgical intervention. But it will mainly deal with the egregious misuses of evolutionary genetics as a justification for repressive social policies, and one of the roads that lead to the Holocaust.

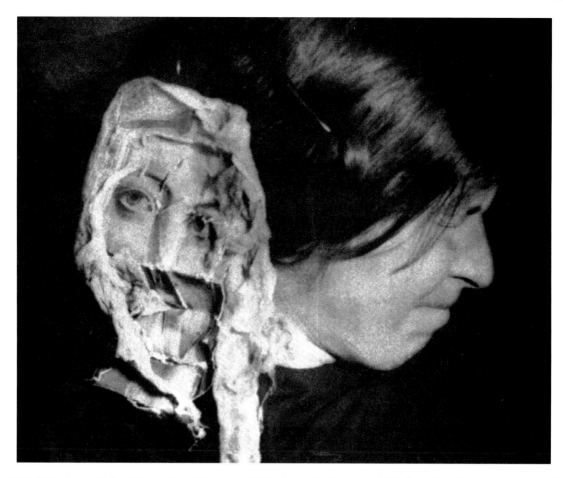

Fig. 19.1. Ectoplasm, Nobel Prize winner Professor Richet's newly discovered life form that, in contrast to endoplasm, materializes outside the body, and can assume human personification. (From Jastrow, 1935.)

GULLIBILITY

"A credulous man: He finds most delight in believing strange things, and the stranger they are the easier they pass with him: but never regards those that are plain and feasible, for every man can believe such" (Samuel Butler, *Characters*, 1667).

That a scientist might be a sharp-minded, skeptical inquirer in his chosen field of study does not necessarily mean that he is not a gullible advocate of others. An amusing case in point is the enthusiastic support of spiritualism by a number of distinguished scientists at the beginning of the twentieth century. From the middle of the nineteenth century, séances had become popular and fashionable. Vic-

torian séances were robust and vigorous, filled with thrilling happenings, such as table turnings, disembodied voice-trumpets, ghostly tunes, and frightful spirit apparitions. If such dramatic manifestations could be confirmed by science, then even the faintest of hearts would be convinced. In the late nineteenth century, photography entered the scene, which, together with scientific controls, promised to provide evidence of the supernatural beyond mere anecdotal description. Many scientists took up the challenge, some of them most eminent.

Alfred Russel Wallace was an early believer, and a life-long enthusiastic advocate. In 1865, he attended a séance in which he heard faint taps, and saw the table vibrated like "the shivering of a living animal" (Kottler, 1974). In other séances, he was convinced that tables were levitated by spirit forces, and, on one occasion, a spirit materialized and brought fresh flowers into the room. Fully convinced of the reality of the spirit world, Wallace became a founding member of the Society for Psychical Research in 1882. He even appeared as an expert witness in 1907, to testify on the validity of spirit manifestations. Wallace was not at all discomforted by the repeated exposure of psychic fraud, declaring that mediums could and naturally would cheat, unless properly controlled, the fault lay with the investigators not taking proper steps to prevent cheating (Kottler, 1974).

Charles Richet (1850–1935) was a distinguished physiologist of outstanding scientific achievement. His pioneering studies on the sometimes fatal shock reaction to injections (anaphylactic shock) opened the science of hypersensitivity, and won him a well-deserved Nobel prize in Medicine in 1913. But he was also a committed evangelist for the spirit world. He ceaselessly advocated his new science, which he called metapsychics, defined in his popular book, *Thirty Years of Psychical Research* as "the science which transcends all physics and biology" (Richet, 1923). Ironically, in spite of his considerable scientific achievements, his enduring popular legacy is in the discredited term "ectoplasm." It was fashionable at that time for mediums to emanate cloud-like substances from the mouth or shoulders, or even from hidden orifices. Sometimes these misty effluvia gradually took on substantial forms, such as persons' faces or isolated hands. After a careful "scientific" examination of the famous medium, "Eva C.," Richet became convinced of the reality of this outside body emanation, and, as a scientist, he proposed that this newly discovered life form was analogous to essential endoplasm and deserved its own appellation: ectoplasm (Fig. 19.1). That Eva C. was later caught in fraud, concealing muslin about her body, which, by a sleight of hand, manifested as ectoplasm, did not dampen Richet's convictions. Like Wallace, he argued that there was always a special explanation for a special lapse.

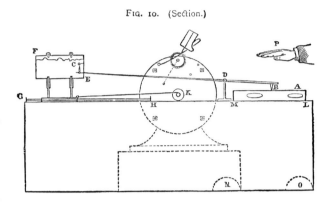

Fɪɢ. 10. (Section.)

Fig. 19.2. Professor Crooke's apparatus for measuring psychic force. One spirit caused an 80-pound displacement. (From Crooke, 1874.)

Fɪɢ. 2.

Fig. 19.3. Crooke's proof of spirit musicality. The accordion suspended in a basket is held by one hand; tunes are played by spirits. (From Crooke, 1874.)

Perhaps the saddest example of an outstanding scientist abandoning common sense to the irrational is provided by Sir William Crooke (1832–1919). As a scientist, Crooke had an impeccable career. He discovered thallium, performed pioneer researches on the cathode ray, invented the Crooke's radiometer, and, for these achievements, became a Fellow of the Royal Society and was awarded a knighthood.

However, conventional science was not sufficient for him. As he declared in his book, *Researches in Spiritualism* (1874), his ambition was to "conclusively establish Psychic Force." He designed complicated arrays of weights and balances to measure psychic pressures (Fig. 19.2), and found one spirit caused an 80 pound displacement. He demonstrated to his own satisfaction, spirit musicality, with an accordion held at one end by a medium, while suspended in a basket (Fig. 19.3), play "Home Sweet Home." He testified to the truth of a floating harmonica, and isolated spirit hands in séances. His most absurd involvement concerned the 18-year-old medium, Florence Cooke and her spirit control, Katie King. When challenged that Florence and Katie were one and the same person, he went into the dark spirit chamber alone and "embraced her," and later "layed his head upon her chest" to hear her heart beat. He was also photographed with Katie arm in arm.

Like Wallace and Richet, Crooke's convictions were not disturbed when his medium, Florence Cooke, was later caught out in fraud. His motivation? Perhaps he just wanted to find proof of an afterlife, or perhaps, as has been suggested by Palfreman (1976), he was driven by an ambition to be the discoverer of a new force, to be "the Newton of psychic science."

RECKLESS INTERVENTION

The great improvements in surgery in the early twentieth century meant that many survived head injuries that previously would have been fatal. World War I, in particular, provided a rich variety of brain-damage cases. The observation that there appeared to be a correlation between the nature of the alterations in a patient's perception and behavior and the site of the brain injury supported the basic assumption of the neurophysiology of that period, that different brain functions have precise anatomical locations. The natural question followed: If brain function was localized in a fine way, could brain malfunction be similarly localized, and can one cure or alleviate brain malfunction by destroying the defective area?

There is a long history of some surgical manipulation on this premise, but a seminal experiment was performed in 1935 by Egas Moniz (1874–1955), chair of neurology at Lisbon University. Moniz was particularly interested in the frontal lobes, which were thought to control social and affective behavior. In 1935, he learned of the experiments of two American physicians, Fulton and Jacobsen, who had completely removed the frontal lobes of the brain (a lobectomy) in two chimpanzees, after which the animals lost the temper tantrums they had been prone to, and became passive.

Audaciously, Moniz and his colleague, Almeida Lima, proceeded to carry out related surgical procedures on human mental-hospital patient in their care. Moniz's initial procedure involved multiple injections of 0.2 mL alcohol into several sites in the prefrontal lobes with the intent of destroying the white nerve fibers (*leukos*), which connect the frontal lobes. But this was too imprecise, and a surgical method was elaborated. The apparatus consisted of a hollow metal tube, the leucotome, which contained a wire loop. Holes were bored in the front of the skull at predetermined sites, and the tube was inserted into the brain to a fixed depth (initially 4.5 cm). The loop was extended, and the offending neuronal pathways were cut.

In 1936, Moniz and Lima reported on their first 20 patients. The results they claimed were very satisfactory: 14 patients were declared cured or improved. Prefrontal leucotomy (or lobotomy, as it later became known) had arrived, and shortly thereafter, neurologist Walter Freeman, professor of Neurology at George Washington University in St. Louis, enthusiastically imported it to the United States.

Moniz advanced a theoretical justification for this irreversible, radically behavior-altering "psychosurgery." He proposed that many mental characteristics, such as creativity, initiative, and anger, were determined by the fine structure of the white matter (the neuronal fibers) of the brain, and that, in healthy people, the pathways were flexible, to produce the behavior appropriate to the occasion

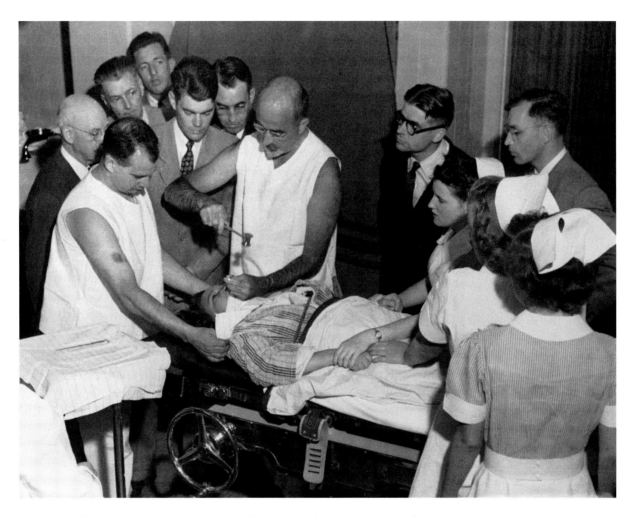

Fig. 19.4. Dr. Freeman demonstrating his ice-pick lobotomy technique. (From Corbis Images.)

(Kucharski, 1984). But this mutability was lost in the mentally ill, who had "abnormally stabilized pathways," and were doomed to continually repeat a behavior, such as rage, no matter what the circumstances. He thought that the frontal lobes were fully involved in certain psychoses, such as schizophrenia and severe paranoia. The logical therapy was, therefore, to cut the abnormal fibers, which would interrupt their endless course, and force a transformation of the pathological thought patterns into more normal ones. There was, in fact, no more scientific evidence for such "abnormally stabilized pathways" than for phrenology, nor was there even any anatomical basis (Kucharski,

1984). Nevertheless, prefrontal lobotomy became a widely adopted treatment for mental illness in other countries. Moniz was awarded the Nobel Prize for Medicine in 1949, for his pioneering work on psychosurgery.

In the United States, Freeman and his colleague, James Watts, a neurosurgeon, elaborated the method into what became known as the "Freeman and Watts standard lobotomy." Here hand-drilled holes were made on either side of the head, to allow the insertion of a blunt knife. Many operations took place under local anesthetic, the advantage being that the patient could be required to perform some mental task, like counting backwards, while cutting proceeded, and the surgeon could tell when he had cut enough (or a little too much) when the patient started to make errors. Dissatisfied with the slowness of the conventional lobotomy technique, Freeman introduced, in 1945, what is, without doubt, the most bizzare surgical procedure in modern medical history: transorbital or ice-pick lobotomy (Fig. 19.4). The point of a stainless steel ice pick was applied to the eye socket bone behind the upper eyelids. A few blows with a mallet sufficed to pierce the thin bone and the pick was driven into the frontal lobes. The pick was then twisted about, destroying lobe tissue. The procedure lasted only 10 minutes, and could be performed under local anesthetic, but this time there was not even a pretense of precision: The surgeon had no idea of the extent of the damage he was inflicting (Vertosick, 1997).

Lobotomy gained prestige and popularity, especially after Moniz was awarded his Nobel Prize. Freeman was elected president of the American Board of Psychiatry and Neurology in 1948. The procedure peaked in the 1950s, then declined. Freeman's last lobotomy was performed in 1967, which ended in the patient's death (daCosta, 1997). The numbers are impressive. It has been estimated that around 40,000 lobotomies were performed in the United States, 17,000 in the United Kingdom, and 9000 in the Scandinavian countries (Tranoy, 1996). The patients were mostly "troublesome" inmates of mental health institutes, of all ages; the youngest in the United States being a 4-year-old child (Kucharski, 1984).

The psychosocial effects of lobotomy were dramatic. Patients often showed increased apathy and passivity, lack of initiative, an inability to concentrate or plan, and no emotional involvement. Clearly, this can hardly be called a cure. Indeed, there is no doubt that the main use of lobotomy was patient control (Tranoy, 1996; Vertosick, 1997). This fact is epitomized in a 1948 report by a leading Norwegian psychiatrist: "Last summer and fall the troublesome women's ward was a real nightmare, and I believe that I have never seen it so bad. Throughout the autumn we took the trouble-makers one after another and had them operated on. Now the ward is completely different" (Tranoy, 1996).

EVOLUTIONARY GENETICS AND SOCIAL REPRESSION

The Problem

As the nineteenth century transformed into the twentieth there was an increasing interest in applying lessons from the biological sciences to society. The core premise was that humans are a part of nature and subject to the laws of nature. But, more particularly, Darwin had made humans the product of natural selection, which he treated in detail in his *The Descent of Man* (1872).

Evolutionary, "Darwinian" interpretations of humans' place in nature quickly appeared. Herbert Spencer coined the phrase "the survival of the fittest," and a picture emerged of a ruthless "nature red in tooth and claw," in which humans were players. Spencer argued that natural selection must be allowed to run its course in human society. But what did this mean? It became widely accepted that many modern "well-meaning" social policies were in fact protecting the weakest and negating the natural selection of the strongest. All right-thinking people had a duty to halt or reverse this degeneration of the human genetic stock. The application of Darwinian principles to human society—social Darwinism—meant reestablishing natural selection by the enactment of policies that would encourage the procreation of the fittest and discourage the spread of the unfit. It was the clear responsibility of knowledgeable and expert scientists to bring attention to the danger, to provide the scientific tools to identify inherited defects and provide scientific solutions for dealing with the problem. The earliest proposals advocated the sterilization of the unfit; extermination of the unfit came later.

In England, there was particular concern over what was perceived to be widening class differences. It was easily shown that the working classes were smaller, weaker, and had more malformations than the upper classes. The military officer or factory owner was taller, and had a much more robust

Fig. 19.5. Eugenics Society poster warning about the dangers of broadcasting bad seed.
(*See* color plate appearing in the insert following p. 82.)

and healthier figure than the common soldier or factory worker. Although it was conceded that environment had something to do with these differences, environmental effects were thought to be minor. Those in charge were so because they had superior inherited traits: They were born to command. The lower classes were born to serve, and the poor were unfit, and should not have been born. Francis Galton (1822–1911) a half-cousin of Darwin, and the coiner of the term "eugenics," advocated the application of scientific breeding to human society. Natural selection must be allowed to continually weed out the most unfit, and any

activity to improve their lot by, for example, providing free food and clothing allowances, was opposed strongly. He argued that, if the lower classes were allowed to continue to propagate inferior children, then they would eventually come to be recognized as enemies of the state and forfeit all claims to kindness (Medewar and Medewar, 1983).

Like Darwin, Galton was a believer in gradualism in inheritance (one of his experiments was to inject the blood of a dark rabbit into a gray pair of rabbits, hoping to produce dappled ones). But, although it is easy to quantify the presence or absence of a heritable trait, how do you quantify variation? Galton was particularly interested in the inheritance of mental traits, and tackled the problem of continuous variation in his influential book, *Hereditary Genius* (1869), in which he presented a normal distributive curve (the "bell curve"), constructed from one million Englishmen distributed into 14 grades of "natural ability" (Provine, 1986). With his colleague, Pearson, he founded modern statistics, and established the prestigious journal *Biometrica*.

It was the responsibility, or more, the duty, of the knowledgeable professional biologist to alert society to the terrible danger of the more fecund lower order outbreeding the better classes. Darwin (1872) quotes the impassioned Mr. Gregg, a colleague of Galton, on this point: "The careless, squalid, unaspiring Irishman multiplies like rabbits; the frugal, foreseeing, self reflecting, ambitious Scot (presumably just the Lowland Presbyterian variety) stern in his morality, spiritual in his faith, sagacious and disciplined in his intelligence ... marries late and leaves few behind him." Eugenic organizations were formed to educate the public about the danger of "bad seeds" (Fig. 19.5). As late as 1929, W. C. Dampier-Whetham, distinguished professor at Cambridge University, complained that the number of children from hereditary peerage marriages had fallen from 7.10–3.13, and saw a similar decline in professional families. "On the other hand Roman Catholic families, miners, unskilled laborers and (much more alarmingly) the feeble-minded were maintaining the numbers of children almost unaltered" (Dampier–Whetham,1929) . If unchecked, he sternly warned, the inevitable result would be the expansion of an "unintelligent proletariat" and the "strains of ability will be weeded out of the nation, with ever increasing danger to civilization" (Dampier-Whetham,1929) . Strong stuff. The eugenics society demanded that only good seed be sown (Fig. 19.5).

The concept that the two classes, workers and their controllers, were becoming genetically more and more separate, and that, eventually, separate

human strains would result, is reflected in Fritz Lang's movie, *Metropolis*, and H. G. Wells 1895 novel, *The Time Machine*.

The alarming discovery that civilization was under threat from the uncontrolled spread of genetic degenerates produced an urgent need to measure human fitness. Measuring people, like measuring animals, became a thriving academic industry. However, while measurements on animals were mainly for the purposes of phylogenetic classification—to distinguish between types of beetle—for humans, the aim was to use body-part measurements to identify mental and behavioral traits, which curiously was neglected for animals. Linking body measurements and character became a major activity in university departments of human science, and journals devoted to human morphometrics flourished. An excellent and amusing account of the absurdities of this "science" is provided by Gould in his, *The Mismeasure of Man* (1981).

But physiogonomy, the science of discriminating character by outward appearance, has a long history. For example, in discussing noses, Aristotle says that people with thick bulbous noses are insensitive; a sharp-tipped nose indicates irascibility; rounded, large, fat noses shows magnanimity; and open nostrils are a sign of passion (*Encyclopedia Britannica,* 11th Edition, 1911). Some scientific explanation of physiogomony was attempted by Darwin, who argues, in the *Expression of Emotions* (1872), that the habitual expression of the muscles of expression may alter the contour of bones by affecting local nutrition. "It is clear," wrote a German professor of genetics in 1941, "that there is a connection between the shape of the nose and personality —a sharp nose being associated with virtue or chastity a blunt one with a sense of humor. The important question is what are the (scientific) principles behind this connection" (Conrad, 1941).

The classification of humans into different races with different (genetic) attributes was almost universally accepted. Just how widespread and commonplace this was illustrated by a 1920 *British*

Medical Encyclopedia entry on "hysteria," which was defined as "a lack of control, tendency to scream and throw uncontrollable fits at the slightest provocation." It noted seriously that, in England, 25 women were afflicted, to each man, but, "in France there are three to four male victims to each hysterical female." Amusingly absurd, but ominous. The same kind of argument was used to justify discrimination of the most brutal type.

In England, eugenic concerns centered around class, but in the United States, a country of immigration, interest concentrated on race and character. One of the most influential of American eugenicists, Charles Davenport, founded and directed the Eugenics Research Organization, which was funded by the Carnage and Rockefeller Foundations. He later became Director of Cold Spring Harbor Laboratory. Davenport maintained that 10% of the U.S. population imposed a heritable burden on the rest, and were a "constant source of danger to the national and racial life" (Reilly, 1987). The reigning school of anthropology was represented by Earnest Hooton, who held that a person's "somatotype" could reveal their intelligence, temperament, moral worth, and probable future achievements. Somatotype could be determined by measuring features such as the relative size of ears, mouth, skull, arms, legs, and hips. An enormous and useless measuring industry was set up. Indeed somatotype photographs of nude individuals of freshmen classes in Vassar, Harvard and Yale were routinely taken, up until the 1950s.

The Immediate Solution

The discovery of the gene in the early twentieth century was seen as a major advance in understanding the mechanisms of hereditary, giving eugenicists the scientific tool for their work. Although it was widely believed in the nineteenth century that inheritable traits were malleable, in the sense that they could be modified, and perhaps improved, by the experience of the parents, the gene of Mendel and Weissman was immutable. Hereditary traits were permanent, and could not be changed. The ines-

capable conclusion was that the only way to save the gene pool of the nation from deterioration was to eliminate the bad genes.

Confidence in the social application of this new science was unbounded, as Pernick (1996) noted, "because they believed their values to be objectively proven, they could dismiss ethical or political criticism as biased, unscientific and therefore irrelevant."

The first task was to identify the defective genes. In the United States, state laws were passed to forbid marriage by alcoholics, epileptics, the retarded, and persons with chronic diseases (Pernick, 1996), but this was quickly regarded as insufficient. Davenport, who even found a gene for thalassophilia, or love of the sea, was an impassioned advocate for sterilization to "dry up the springs that feed the torrent of defective and degenerate protoplasm" (quoted in Friedlander, 1995). Eligible categories were extended to include epileptics, habitual criminals, rapists, or the insane (a widely defined category, which could include the uncontrollably promiscuous).

The first compulsory sterilization law was passed in Indiana in 1907, and such measures were eventually adopted in 30 states. By 1963 (when most sterilization programs had stopped), more than 60,000 U. S. citizens had been involuntarily sterilized (Reilly, 1987). In the 1930s, the success of eugenic sterilization in the United States encouraged other countries such as Canada, France, Norway, and Denmark to adopt such laws.

The Final Solution
PRE-NAZI GERMANY

Eugenics had wide support in the academic and medical communities of early twentieth century Germany. A seminal text written by the biologist, Alfred Ploetz in 1895, *Grundlinien einer Rassenhygiene*, argued that traditional medical care helps the individual, but endangers the race. A new kind of hygiene was called for: racial hygiene (*Rassenhygiene*). This concept caught on, and by the 1930s, courses in *Rassenhygiene* had been established in most German universities.

Questionnaire 1
Case no. ....
          Name of Institution: ....
          in: ....
First and family name of patient: .... maiden name: ....
Date of birth:.......... City: .... District: ....
Last residence:.......................... District: ....
Unmarr., marr., wid., div.: Relig:.. Race^a .... Natlty: ....
Address of nearest relative: ....
....
Regular visits and by whom (address): ....
....
Guardian or Care-Giver (name, address): ....
....
Cost-bearer:.... How long in this inst.: ....
In other institutions, when and how long: ....
How long sick:.. From where and when transferred: ....
Twin $^{yes}_{no}$ ... Mentally ill blood relatives: ....
Diagnosis: ....
....
Primary symptoms: ....
....
Mainly bedridden? $^{yes}_{no}$... Very restless? $^{yes}_{no}$..... Confined? $^{yes}_{no}$ ....
Incurable phys. illness: $^{yes}_{no}$...... War casualty: $^{yes}_{no}$......................
          For schizophrenia: Recent case ...... Final stage .. good remission. ...
          For retardation: Debility: .... Imbecile:.... Idiot: ....
          For epilepsy: Psych. changes........ Average freq. of attacks ....
          For senile disorders: Very confused .... Soils self ....
Therapy (Insulin, Cardiazol, Malaria, Salvarsan, etc.):. Lasting effect: $^{yes}_{no}$......
Referred on the basis of §51, §42b Crim. Code, etc. .... By ....
Crime: ... Earlier criminal acts: ....
Type of Occupation: (Most exact description o´ work and *productivity,* e.g. Fieldwork,
does not do much.—Locksmith's shop, good skilled worker.—No vague answers, such
as housework, rather precise: cleaning room, etc. Always indicate also, whether con-
stantly, frequently or only occasionally occupied)
....
....
Release expected soon:..

    ^{a}German or related blood (German-blooded), Jew, Jewish *Mischling* [half-breed] 1st
or 2nd degree, Negro *(Mischling),* Gypsy *(Mischling),* etc.

Do not mark in this space.

..... Place, Date ....

....
(Signature of medical director or his representative)

INSTRUCTION SHEET
To be followed in filling out the questionnaires
All patients are to be reported who
1. suffer from the diseases enumerated below and who within the institution
   can be occupied not at all or only at the most mechanical work
   (picking, etc.):
   Schizophrenia,
   Epilepsy (indicate if exogenous, war-related or other causes),
   Senile disorders,
   Therapy-resistant paralysis and other Lues [syphilitic] diseases,
   Retardation from whatever cause,
   Encephalitis,
   Huntington's chorea and other terminal neurological conditions;
   or
2. have been continuously in institutions for at least 5 years;
   or
3. are in custody as criminally insane;
   or
4. do not possess German citizenship or are not of German or related
   blood, giving/designating race^b and nationality.
The questionnaires, to be filled out individually for each patient, are to be given serial
numbers.
The questionnaires are to be filled out by typewriter whenever possible.
Due on ....
In the case of patients sent to this institution from outside the evacuation area, a (V)
is to be placed behind the name.
In case the number of Questionnaire 1 forms sent are not sufficient, please order the
number needed through my office.

    ^{b}German or related blood (German-blooded), Jew, Jewish *Mischling* 1st or 2nd class,
Negro, Negro *Mischling,* Gypsy, Gypsy *Mischling,* etc.

SOURCE: Questionnaire translated from Judgment in Hadamar Trial, Frankfurt/M., February–
March 1947 (4 KLS 7/47), Landgericht Frankfurt. Instruction sheet from Heyde Trial Docu-
ments, pp. 210–11. Questionnaire and instruction sheet translated by Amy Hackett.

Fig. 19.6. English translation of Nazi Hereditary Health Court Questionnaire. (From Lifton, 1986.)

In 1921, an influential two-volume book was published, *Grundriss der Menschlichen Erblich-keitslehre und Rassenhygiene* (*Outline of Human Genetics and Racial Hygiene*), by Erwin Bauer, Eugene Fischer, and Fritz Lenz. It became a highly regarded textbook, and went into many editions and translations, including an English version entitled *Human Heredity,* published in 1931. But, although the book appealed to scientific neutrality, it uncompromisingly declared the case for the genetic differences between primitive and advanced races, and the genetic base for the superior accomplishments of the Nordic race. In the words of Proctor (1988), Lenz's racial science was "soaked with the values and prejudices of the times." It was read by Hitler while in Landsberg prison, after the failed 1923 Munich putsch, and was much quoted by him in support of his racial policies.

Developing the theme, Lenz argued that genetic degeneracy had gone so far that immediate intervention was called for. He advocated that, in order to purify the race, it is necessary to ruthlessly cull the contaminated by sterilizing people with the slightest suggestion of mental disease with the exception of the artistically gifted. He reckoned that this would involve almost one-third of the German people (Proctor, 1988).

NAZI BIOLOGY

Hitler made biology, and race hygiene in particular, a center plank in his platform, asserting "the primacy of national biology over national economy," and demanding "the most modern of medical means" to achieve these goals (Hitler, 1923). He claimed that national socialism was simply applied biology, whose goal was to make a healthy race and protect against race pollution. This leading role for their discipline was widely embraced by university biologists, and many a German physicians rushed to qualify as a "genetics doctors" (*Erbarzt*). In fact, medical doctors joined the Nazi party in greater numbers than any other professional group. By 1936, 40% of Berlin physicians under 40 years of age, and who were eligible to join, had become members of the National Socialist (NS) Physicians League (Proctor, 1988). Funding for Biology from the state research funding agencies, the Reichsforschungsrat (RFR) and the Deutsche Forschungsgemeinshaft (DFG) rose dramatically, from about 50,000 RM in 1934 to about 500,000 RM in 1941—genetics and mutation research accounting for 50% of the total (Deichman, 1996). The SS even had its own genetic research institute, Das Ahnenerbe. The great tragedy of "racial cleansing" which followed, culminating in the horrors of the Holocaust, is unique in using science—perverted science—to justify its evil.

Sterilization. One of the first laws passed by the Nazis legalized compulsory sterilization of the handicapped and those with hereditary disease. In 1933, special Hereditary Health Courts (*Erbgesundheitsgerichte*) were set up; 1700 genetic health courts were planned. The court panel consisted of three experts—purportedly impartial and scientific—two of whom were physicians (one specializing in public health, one in genetics), and one of whom was a judge. Their task was to examine suspects for "hereditary sickness" (*Erbkrankheiten*), and individuals found to be tainted were legally ordered to be sterilized. There was even an appellate genetic health court (*Erbgesundheitsobergericht*) to handle any appeals, although few were made, and fewer were successful. The categories for compulsory sterilization initially included the mentally feeble, schizophrenics, epileptics, and the congenitally blind or deaf, but was later broadened to include people with antisocial behavioral traits, such as alcoholism. Ironically, at the same time, abortion was designated "racial treason" (*Rassenverrat*) and was made a capital offense, and live animal experimentation was severely restricted.

Initially, male victims were sterilized by vasectomy and females by tubular ligation. But this was found too time-consuming: Males were then castrated, and the pelvic region of women was exposed to high levels of X-rays. It has been estimated that, by 1939, as many as 400,000 Germans were sterilized under these laws, or 5% of the population (Müller-Hill, 1998), which required an enormous logistical effort.

Euthanasia and Extermination. Sterilization was soon found to be too "wasteful," because it produced large numbers of "useless" people to house and feed. Euthanasia was advocated as the "efficient" solution. Again, "science" was used to support this step, most importantly in a 1920 book by Karl Binding, a lawyer, and Alfred Hoche, a psychiatrist and neuropathologist, called *Die Freigabe der Vernichtung lebensunwerten Lebens* (*Permission for the Destruction of Life Unworthy of Life.*) It introduced the concept of a life not worth living (*lebensunwerten Lebens*), and argued that the law should permit the killing of such people, who were only empty shells of humans. However, the killing had to be painless, and could only be administered by an expert. Their solution was not unique: The French Nobel Prize winner for medicine, Alexis Carrel, had suggested in his 1935 book *Man the Unknown*, that the criminal and the insane should be "humanely and economically disposed of in small euthanasia institutions supplied with proper gasses."

A killing law was issued, as soon as war broke out in 1939, by a top-secret directive from the high office of the Chancellery of the Führer, ordering the immediate euthanasia of "hereditarily defective" children. The machinery to kill the handicapped was already in place, and the killings began about October 1939. Doctors and midwifes were required to report all infants with gross physical deformities or congenital abnormalities, and to fill out a questionnaire. An expert panel examined and marked the questionnaires, a positive (+) indicating that the child should be killed, a minus (−) that it should be excluded from the program. The law initially covered children up to 3 years old, but by

1941, it was extended to adolescents up to 17 years old. Rapid, large-scale, efficient killing required organization, and top-secret killing centers were set up. The most important, in Görden Hospital near Berlin, was equipped with special killing wards, which served as training centers in the techniques of extermination (Lifton, 1986). Initially, children were killed by starvation, but a barbiturate overdose was found to be more rapid and efficient. After death, the parents were sent a condolence letter and a death certificate with a false cause of death. It has been estimated that more than 5000 children were killed in this program by 1941 (Friedlander, 1995).

With the outbreak of war, tightened party control was also seized on to implement a top-secret program for euthanizing institutionalized adults, code named T4 (Teirgarten 4). Again, using a front of scientific and legal objectivity, the Chancellery of the Führers required an official registration form to be filled out for all the inmates, with the purpose of identifying the incurably ill and the hereditarily diseased (Fig. 19.6). As detailed in Friedlander (1995), instructions were supplied, specifying who had to be reported:

1. Patients institutionalized for five or more years.
2. Patients with the following conditions, if they were unable to do work in the institution, or could only do routine labor.
   a. Schizophrenia
   b. Epilepsy
   c. Senile diseases
   d. Therapy-resistant paralysis and syphilis
   e. Encephalitis
   f. Huntington's disease and other terminal neurological diseases
3. Patients committed as criminally insane.
4. Patients without German citizenship.
5. Patients not of Germanic or related blood (this category included Jews and Gypsies).

To make the process appear scientific, legal, and impartial, an expert panel adjudicated who was to die, and, if appealed, there could be a final review by a chief expert or a higher court. Auto-matic extermination orders passed on those categorized as "life-without-living" and on "useless eaters." Jewish inmates were condemned under category 5 by itself. The initial plan was to euthanize 0.1% of the German population, i.e., about 70,000, and, by August 1941, when Hitler ordered gassing of inmates of psychiatric units to be halted, 70,000 had been killed, on schedule. This goal was achieved by improving, making more effective, the killing process. Adults marked (+) were gathered together and transported in special SS-disguised trucks to the killing centers. Initially, overdoses of sedatives were used, but gassing was found to be more efficient. Carbon monoxide was released from tanks into specially built gas chambers, killing many quickly. False (but legal) death certificates were issued, usually noting TB as the cause of death.

Academics were intimately involved at all levels. For example, in 1940, Omar von Verschuer, Director of the Institute of Hereditary Biology and Race-hygiene at the University of Frankfurt, called for the setting up of registers of all "social misfits" in Germany, "so that we may combat asocial mentality with all the means at our disposal"(Müller-Hill, 1998). The categories of people to be exterminated was greatly expanded after 1941, in what as been called "wild euthanasia," untrammeled by the previous "legalities" (Lifton, 1986). Only inmates of mental institutions, but the population at large was vulnerable, including people with tuberculosis, deviant political attitudes (communists), deviant sexual attitudes (homosexuals), and asocial behavior (alcoholics and drug addicts). The Jewish race was identified as a diseased race, or "race poison," and was targeted for extermination (Reitlinger, 1961). Desperately, many "half-Jews," who had a German mother and a Jewish father, invented the fiction that their Jewish father was not their real father. Hundreds of such claims were investigated by professional geneticists, who later claimed that the fate of persons was not their business, that they just solved the paternity question scientifically (Müller-Hill, 1993). In the conquered territories outside the borders of Germany, there was no judicial limitation to implementing the mass annihilation, which took place on a scale unprecedented in human history.

Genetic Experiments. The Third Reich provided lavish funding to investigate the heritability of human disease, through the state research-granting agency, the DFG (Proctor, 1988). There was no lack of applicants from universities and hospitals. Twin studies were particularly favored. In a report to the DFG, Professor von Verschuer wrote:

> "My assistant Dr. Mengele has joined me in this branch of research. He is presently employed as Hauptsturmführer and camp physician in the concentration camp of Auschwitz. Anthropological investigations on the most diverse racial groups of this concentration camp are being carried out ... The blood samples are being sent to my laboratory for analyses" (Proctor, 1988).

Convicted war criminal, Dr. Joseph Mengele, was infamous for his cruel twin experiments, such as injecting twins with typhus.

A director of the Görden hospital, Hans Heinze, supervised the training of physicians at the Görden killing center. Autopsies were made on all bodies and a collection of tissue samples was carefully maintained to satisfy requests by university researchers. There were studies on the effects of hypoxia and decompression on "hereditary" epileptics (Diechman, 1996). There was a special interest in studying the relationship between brain structure and mental malfunction, and a program was set up at Görden, in which selected patients (such as the mentally retarded or epileptics) were observed, given specific tests, then killed. Their brains were quickly removed and sent to the involved experts. Between 1940 and 1942, the well-known neuroanatomist, Julius Hallervorden, Director of the Kaiser Wilhelm Institute for Brain Research, received 697 brains of euthanized persons, for examination, many of them from his own patients. In his post-war interrogation, he justified his behavior by arguing, that since the people were going to be killed, why waste the brains (Müller-Hill, 1998).

Who took part in this abomination? Lifton (1986) suggests that it was a matter of medicalized killing, and that the motives of the physicians were ideological. Friedlander (1995) argues that it involved the self-selection of murderers and sadists in the medical and biological professions.

20 Today

What is truth? said jesting Pilate; and would not stay for an answer.
(Francis Bacon, *Essay On Truth*, 1625.)

WHAT IS SCIENCE?

The theme that this book attempts to develop is that the " what and how" of scientific enquiry has been developed and honed over its long history, and that science has been able, particularly since the seventeenth century, to provide more and more accurate, truthful descriptions of nature and its workings. Well into the eighteenth century, many scientists believed, or at least asserted, that they were moved to reveal God's great design in nature, to show the beautiful intricacy of the universal harmony, but in nineteenth century science, God was dropped.

The conviction that scientists were providing empirical, rational, and testable explanations of the universe, which become increasingly correct as new discoveries are made, became a prime motivation for their work and a source of intellectual pleasure. The break with other types of explanation is that, for example, physiology deals with body function in terms of measurable physics, chemistry, and mathematics, but mythological accounts are simply a matter of belief, complete in themselves. As an illustration of this distinction between scientific and other types of explanation of natural phenomena—albeit a rather forced one—consider the proposition that a horse had been hatched from a hen's egg. In the realm of magic, myth, and religion, why not? All is possible, there are no bounds, the rules of the universe are not deducible through reason, but are instead simply revealed. Now, the ancient Greeks would have been skeptical, arguing that such an event is against natural law, encapsulated in phrases such as "like begets like." In premodern times, this type of event would have been considered very unlikely, but possible, perhaps as a portent of dreadful happenings. For example in *Macbeth*, on the night of Duncan's murder, Duncan's horses eat each other (Ross: "They did so, to the amazement of mine eyes"). Today, such an event is considered widely inconsistent with the known scientific, materialist *mechanisms* of reproduction, development and animal behavior.

179

BIOLOGY TODAY

Promises

By the early twentieth century, the scientific method of investigation had matured: To a great extent, further developments are variations on a theme. The consequent rapidly growing understanding of biological processes brings with it the possibility, indeed the certainty, of increasing control over life processes. Today, there is an audacious, perhaps premature, attempt to functionally integrate biology at all levels, from the molecule to cell, to organism, to community, to ecology—a sort of scientific great chain of being. But, more importantly, the recent advances in genomics, integrating gene expression into physiology, point to a level of biological manipulation undreamed of before. Techniques such as gene identification and gene manipulation are becoming commonplace. Perhaps the twentieth century was the century of physics, but the twenty-first promises to be one of biology. The promise is great, and will require many important choices and decisions.

In agriculture, there is an enormous amount of genetic research to modify farm animals for desirable traits, such as increased yields of milk or meat, and to genetically engineer plants to express pesticides or become resistant to specific herbicides.

In human biology, genetic fingerprinting is making selective abortion possible on an unprecedented scale. Not only can we now detect and abort fetuses with serious genetic handicaps, such as Down's syndrome or Huntington's chorea, but it will soon be possible to "select" for desirable cosmetic qualities, such as sex, height, color of hair, athletic scope, and so on.

The new emerging gene therapy will revolutionize health care. The goal of gene therapy is to protect against, alleviate, or repair damaged tissue from such insults as stroke, through, for example, genetic manipulation to increase the targeted tissue's vascularization. It is very likely that, in the near future, it will be practical to combat inborn genetic faults by adding corrective genes to targeted cells with defective genes, using biological, chemical, or electronic means. There is currently much research into methods for repairing defective genes, so that the corrected gene would be expressed at the right time to produce the missing protein in normal amounts, and thus restore normal function.

Stem cell research and cloning offers the possibility of producing new immunocompatible replacement organs. In this technique, a nucleus from a nonreproductive cell is taken from a patient's body and fused with a donated egg cell (oocyte). The objective is that the resultant embryo can serve as a source of stem cells genetically identical with the patient. In principle, these can be used to replace diseased organs, such as the liver or kidney. For example, embryonically derived brain cells might be grafted onto the brain to cure diseases caused by loss of specific types of neurons, such as the dopamine-producing neurons lacking in Parkinson's disease patients, or the missing oligodendrocytes in multiple sclerosis patients.

Problems

The new molecular biology is also producing new problems, some of which are so novel that we have little in the way of legal precedent or cultural heritage to deal with them. As was discussed throughout this book, the practice of science at any particular time is inextricably bound up with the prevailing ideological climate, and, as illustrated in Chapter 19, society has shown no reluctance to use science for the benefit of dogma or profit.

There is the question of ownership. As J. D. Bernal (1965) said, science must be funded, it must answer to its paymasters. Who are the paymasters of these new, revolutionary genetic techniques? Who owns them, the public, or private industry? Can, or should, for example, the gene sequence for hemophilia be owned by a private company?

The more accurate genetics raises the specter of neosocial Darwinism, and is all the more dangerous, because it is more "scientific," and has its emphasis, more neutrally, on the individual, rather than the race. Parents will be able to abort children that have genetic handicaps for unwanted health, mental, or behavioral traits, or who do not have desired features, such as size, sex, athleticism, or color of hair. This question of social control through genetic tests arises anew. Will a new underclass of people with "flawed " genes be created, discriminated against in employment or insurance? In October 2000, the British Labour government authorized the use of genetic testing for insurance coverage.

There is the question of human worth. The use of the human fetus as a source of replacement tissue, as a commodity, raises uncomfortable questions about when the developing child has human rights, and at what stage in development is the threshold to disallow killing for genetic material. Such research is strongly condemned by the Vatican, as well as many ethicists and scientists. The use of living human fetal tissue for research is currently illegal in Germany, and has been disapproved by the European Union ethics committee (Dickson, 2000). In the United States, federal funds currently cannot be used to produce new human embryo stem cell research, although the National Institutes of Health proposes to allow research on existing stem cells, but not their extraction. Ironically, the opposite view is advocated by Peter Singer, professor at Princeton University. Singer, the philosophical father of the animal rights movement, who holds that "speciesism" (discrimination against animals by humans) is as deplorable as sexism and racism, argues that a newborn baby does not have human consciousness and the same claim to life as a person, and that it should be permissible to kill the child up until it is one month old, if this "increases the total amount of the family's happiness" (Singer, 1979).

Reaction

Perhaps fueled by fear of a "brave new world," the end of the twentieth century saw a revival of antiscience irrationalism. In the last 20 years, "postmodernism" has taken root in many university departments of sociology and philosophy. The basic premise of postmodernism is that science is just a reflection of modern belief systems, that it is culturally determined, and is no more

valid or true than any other culturally determined explanation of the universe. Science is accused of being sexist, authoritarian, monocultural, imperialist, capitalist— all in all, not "politically correct." Scientists are seen as the priests of this belief system, presiding over the justifying ritual in the new temple, the laboratory. In this brave new "politically correct" world, everyone has the right to believe what they like, and all beliefs are equally valid. For example, Stephen Clark, professor of philosophy at Liverpool University, argues that the celestial "theories" of the Tewa Indians have "worked … as well as those of the physicists next door in Los Alamos," and asks "can any of them be true?" (Clarke, 1995). For Trevor Pinch of Cornell University, "scientists are (simply) skilled experts like any other skilled experts, such as realtors, plumbers and chefs," and suggests that "different styles of science may possess a gender valence" (Pinch, 1996). The language of this discourse is typically obscure.

In 1996 an annoyed physicist, Alan Sokal, submitted a paper on the sociopolitics of quantum gravity to the leading postmodern journal, *Social Text*. It was an elaborate spoof or parody of postmodernistic thinking, with a meaningless title, "Transgressing the boundaries: towards a transformative hermeneutics of quantum theory." The paper was filled with empty and meaningless jargon and irrelevant equations, and concluded that "physical reality … is at bottom a social and linguistic construct." As detailed in their book, *Fashionable Nonsense, Postmodern Intellectuals' Abuse of Science* (Sokal and Brichmont, 1998), this provocative hoax sparked a furious debate, which still continues. There would be something ridiculous about a professor, who had a triple bypass, sitting on an airplane flying over the Atlantic, writing on his lap-top computer that the theory that the heart is a pump powering the circulation of the blood is a culturally determined myth no more valid than that of traditional Viking or Amerindian accounts of the evil eye. In the end, the basic silliness of the neomodernists, as

Paracelsus might have said, amounts to philosophically voiding into the wind.

Conclusion

Unless there is a collapse of our civilization, science is neither reversible nor stoppable. If science history teaches any lesson, it is that, at an accelerating pace, new and completely unpredicted discoveries will be made in the future with unimagined results. In the second edition of his book on the future of science, J.D. Bernal (1969) declared that molecular biology was the greatest and most comprehensive idea in all science. Early signs indicate that his prediction was correct, and that the twenty-first century will be radically influenced by the powerful combination of bio-molecular discoveries and computational advances. The challenge will be to use and control the new knowledge. Another lesson in the history of science however, is that the increasing power to do good carries with it the threat of even greater harm. Like never before, it is imperative that citizens and politicians become better educated in the methods and goals of science—past and present. Scientists themselves should not hesitate to contribute to this general education.

References

Agassiz, L. and A. A. Gould. 1851. *Outlines of Comparative Physiology, Touching the Structure and Development of the Races of Animals, Living and Extinct* London: H. G. Bohn.

Aldhous, P. 2000. Cloning owners go to war. *Nature*, 405: 610–612.

Ashton, J. 1883. *Social Life in the Reign of Queen Anne: Taken from Original Sources*. London: Chatto and Windus.

Ball, J. 1928. *The Sack-'Em-Up Men*. Edinburgh: Oliver and Boyd.

Balter, M. 1999. New light on the oldest art. *Science* 2893: 920–922.

Bartholin, T. 1653. *Vasa lymphatica, nuper hafniae in animantibus inventa, et hepatis exequiae*. Paris: Du Paris.

Bastholm, E. 1950. *A History of Muscle Physiology*. Copenhagen: Munksgaard (Thesis).

Bauer, E., E. Fischer, and F. Lenz. 1921. *Grundriss der munschlichen Auslese und Rassenhygiene*. Munich: J. F. Lehmans.

Baumer, A. 1987. Zum Verhaltnis von Religion und Zoologie im 17. Jahrhundert (William Harvey, Nathaniel Highmore, Jan Swammerdam). *Ber. Wissenschaftgesch*. 10: 69–81.

Beaumont, W. 1833. *Experiments and Observations on the Gastric Juice and the Physiology of Digestion*, 1959 edition. New York: Dover.

Behe, M. J. 1995. *Darwins Black Box: The Biochemical Challenge to Evolution*. New York: Free Press.

Bernal, J. D. 1965. *Science in History: The Emergence of Science*, 3rd ed. Cambridge, MA: MIT Press.

Bernal, J. D. 1969. *The World, the Flesh and the Evil*. Bloomington: Indiana University Press.

Bernard, C. 1878. *Phenomena of Life Common to Animals and to Plants*, 1974 edition. Dundee: Burns and Harris.

Bernard, C. 1927. *An Introduction to the Study of Experimental Medicine*. 1865 Macmillan.

Binding, K. and A. Hoche. 1920. *Die Freigabe der Vernichtung lebensunwerten Lebens*. F. Meiner.

Bittar, E. E. 1956. The influence of Ibn Nafis: a linkage in medical history. *Univ. Mich. Med. Bull*. 22.

Bohr, C., K. Hasselbalch, and K. Krogh. 1904. Über einen in biologischer Beziehung wichtigen Einfluss, den die Kohlensaurespannung des Blutes auf dessen Sauerstoffbindung übt. *Skand. Arch. Physiol*. 16: 402–412.

Borelli, G. A. 1734. *De motu animalium. Part I and II*. Neapoli: Typis Felicis Mosca.

Boring, E. 1957. *A History of Experimental Psychology*, 2nd ed. New York: Appleton-Century-Crofts.

Bower, B. 1966. Visions on the rocks. *Sci. News* 150: 209–224.

Bowman, W. 1842. On the structure and use of the Malpighian bodies of the kidney, with observations on the circulation through the gland. *Philos. Trans. Roy. Soc*. 132: 57.

Boylan, M. 1982. The digestive and "circulatory" systems in Aristotle's biology. *J. Hist. Biol*. 15: 89–118.

Breasted, J. 1930. *The Edwin Smith Surgical Papyrus*. Chicago: University of Chicago Press.

Brooke, D. 1930. *Private Letters Pagans and Christians*. New York: Dutton.

Burton, R. 1938. *The Anatomy of Melancholy (1628)* (translators F. Dell and P. Jordan-Smith). New York: Tudor.

Cahill, T. 1995. *How the Irish Saved Civilization*. New York: Doubleday

Carrel, A. 1935. *Man the Unknown*.

Castiglioni, A. 1958. *A History of Medicine,* 2nd ed. New York: Knopf.

Celsus, A. 1935. *Deremedicina. (English translation by W. G. Spencer)*. Cambridge, MA: Harvard University Press.

Chauvet, J-M., E. Brunel Deschampres, and C. Hillaire. 1996. *The dawn of art: The Chauvet cave*. New York: Abrams.

Choulant, L. 1920. *History and Bibliography of Anatomical Illustration*. Chicago: University of Chicago Press.

Clagett. 1994. *Greek Science in Antiquity*. New York: Barnes and Noble.

Clarke, S. R. L. 1995. A map of everything. in *New York Times Book Review*: 34–34.

Conrad, K. 1941. *Der Konstitutionstypus als genetisches Problem*. Berlin: Springer Verland.

Contadini, A. 1994. The Ibn-Buhtisu bestiary tradition. *J. Hist. Med*. 6: 349–364.

Cournand, A. 1982. Air and blood in: A. P. Fishman and D. W. Richards, eds., *Circulation of the Blood: Men and Ideas,* pp3–70. Bethesda, MD: American Physiological Society.

Crombie, A. C. 1959. *Medieval and Early Modern Science*. Garden City, NY: Doubleday.

Crookes, W. 1874. *Researches in spiritualism*. London: James Burns.

Cross, S. 1981. John Hunter, the animal economy, and late eighteenth-century physiological discourse. *Stud. Hist. Biol*. 5: 1–110.

Culotta, C. 1970. Tissue oxidation and theoretical physiology: Bernard, Ludwig, and Pfluger. *Bull. Hist. Med*. 44: 109–140.

da Costa, D. A. 1997. The role of psychosurgery in the treatment of selected cases of refactory schizophrenia: a reaprpraisal. *Schizo. Res*. 28: 223–230.

Dampier-Whetham 1929. *History of Science*. Cambridge, MA: Cambridge University Press.

Darwin, C. 1859. *The Origin of the Species by Means of Natural Selection or the Preservation of Favored Races in the Struggle for Life*. London: John Murray.

Deichmann, U. 1996. *Biologists under Hitler*. Cambridge, MA: Harvard University Press.

Diamond, J. 1993. Evolutionary physiology: pp89–112. *The Logic of Life*. Oxford: Oxford University Press.

Dickson, D. 2000. European panel rejects creation of human embryos for research. *Nature* 408: 227.

Dijkgraaf, 1960. Spallanzani's unpublished experiments. *ISIS* 51: 9–20.

Edsall, J. 1972. Blood and hemoglobin: the evolution of knowledge of functional adaptation in a biochemical, part 1: The adaptation of chemical structure to function in hemoglobin. *J. Hist. Med.* 5: 205–257.

Eknoyan, G. 1996. Sir William Bowman: His contributions to Physiology and nephrology. *Kidney NT.*, 50: 2120–2128.

Eknoyan, G. 1999. Santorio Sanctorius (1561–1636): founding father of metabolic balance studies. *Am. J. Nephrol.*19: 226–233.

Eldon. 1965. *History of Biology*. Minneapolis: Burgess.

Farrington, B. 1944. *Greek Science*. Harmondsworth, UK: Penguin.

Farrington, B. 1949. *Greek Science*. Harmondsworth, UK: Penguin.

Fine, L. 1987. British contributions to renal physiology: of dynasties and Diuresis. *Am. J. Nephrol.* 19: 257–265.

Fine, L. G. 1887. Evolution of renal physiology from earliest times to William Bowman: C. W. Gottschalk, R. W. Berliner, and G. H. Giebish, eds. *Renal Physiology: People and Ideas*. Bethesda, MD: American Physiological Society, pp. 1–30.

Fishman, R. 1997. The origin of the species, man's place in nature and the naming of calcarine sulcus. *Doc. Ophthal.* 94: 101–111.

Fleming, D. 1955. William Harvey and the pulmonary circulation. *ISIS* 46:64–69.

Fort, G. 1883. *Medical Economy During the Middle Ages*. New York: JW Bouton.

Foster, M. 1877. *Textbook of Physiology*, London.

Foster, M. 1901. *Lectures on the History of Physiology During the 16th, 17th and 18th Centuries*. London: Cambridge University Press.

Frampton, M. F. 1991. Aristotle's cardiocentric model of animal locomotion. *J. Hist. Biol.* 24: 291–330.

Frank, R. G. 1980. *Harvey and the Oxford Physiologists: A Study of Scientific Ideas*. Berkley: University of California Press.

French, R. 1978a. The thorax in history: From ancient times to Aristotle. *Thorax* 33: 10–18.

French, R. 1978b. The thorax in history: Hellenistic experiment and human dissection. *Thorax* 33: 153–166.

Frey, E. 1975. The earliest medical texts. *Med. J. Aust.* 2: 949–952.

Friedlander, H. 1995. *The Origins of Nazi Genocide: From Euthansia to the Final Solution*. Chapel Hill: University of North Carolina Press.

Fruton, J. S. 1976. The emergence of biochemistry. *Science* 192: 327–333.

Fulton, J. F. 1953. *Michael Servetus, Humanist and Martyr*. New York: Reichner.

Galton, F. 1869. *Hereditary Genius: An Inquiry into Its Laws and Consequences*. London: MacMillan.

Gardner, E. 1965. *History of Biology*. Minneapolis: Burgess.

Garland, T. J. 1994. Evolutionary physiology. *Annu. Rev. Physiol.* 56: 579–621.

Gatrel, V. A. C. 1994. *The Hanging Tree: Execution and the English People 1770–1868*. London: Oxford University Press.

Geison. 1978. *Michael Foster and the Cambridge School of Physiology: The Scientific Enterprise in Late Victorian Society*. Princeton: Princeton University Press.

Gerszten, P., E. Gerszten, and M. Allison. 1998. Diseases of the skull in pre-Columbian South American mummies. *Neurosurgery* 42: 1145–1151.

Goldsmith, O. 1774. *A History of the Earth and Animated Nature*.

Gottschalk, C. W., R. W. Berliner, and G. H. Giebisch eds., 1987. *Renal Physiology: People and Ideas*. Bethesda, MD: American Physiological Society.

Gould, S. J. 1981. *The Mismeasure of Man*. New York: Norton.

Grafflin, 1928. The structure and function of the kidney of Lophius piscatorius. *Bull. Johns Hopkins Hosp.*, 43: 205.

Griffin, D. 1959. *Echos of Bats and Men* New York: Doubleday.

Gross, C. 1999. *Brain, Vision, Memory: Tales in the History of Neuroscience*. Cambridge, MA: MIT Press.

Gruner, O. C. 1930. *A Treatise on The Canon of Medicine of Avicenna*. London: Luzac and Co.

Grüsser, O.-J. and M. Hagner. 1990. On the history of deformation phosphenes and the idea of internal light generated in the eye for the purpose of vision. *Doc. Ophthal.* 74: 57–85.

Guerlac, H. 1997. *Essays and Papers in the History of Modern Science*. Baltimore: Johns Hopkins.

Haas, L. 1997. Jean Baptiste van Helmont (1577–1644). *J. Neurol. Neurosurg. Psychiatry.* 63: 428.

Haggard, H. W. 1928. *Devils, Drugs and Doctors*. New York: Harper Row.

Hales, S. 1740. *Statistical Essays*, 2nd ed. London: Innys.

Hall, T. S. 1969a. *History of General Physiology. From Presocratic Times to the Enlightenment*. Chicago: University of Chicago Press.

Hall, T. S. 1969b. *History of General Physiology: From the Enlightenment to the End of the Nineteenth Century*. Chicago: University of Chicago Press.

Hall, T. S. 1972. *Treatise on Man: René Descartes: French Text with Translation and Commentary*. Cambridge, MA: Harvard University Press.

Hall, W. 1987. Stephen Hales: theologian, botanist, physiologist, discoverer of hemodynamics. *Clin. Cardiol.* 10: 487–489.

Haller, A. v. 1754. *Physiology-Course of Lectures upon the Visceral Anatomy and Vital (Economy of Human Bodies-Compiled for the Use of the University of Gottingen)*. London: Innys and Richardson.

Haller, A. v. 1786. *First Lines of Physiology*. Edinburgh: Elliot.

Hamilton, W. F. and D. W. Richards. 1982. The output of the heart. Pages in A. P. Fishman and D. W. Richards, eds., *Circulation of the Blood: Men and Ideas*. Bethesda, MD: American Physiological Society.

Harvey, J. E. 1957. *Movement of the Heart and Blood in Animals. An Anatomical Essay*. Oxford: Blackwell.

Hatfield, G. C. and W. Epstein. 1997. The sensory core and the medieval foundations of early modern perceptual theory. *ISIS*, 70: 363–378.

Hess, C. 1994. Entwicklung der Neurophysiologie im 19 Jahrhundert Schweiz. *Rundschau Med.* 83: 483–490.

Hierholzer, K. and K. J. Ullrich. 1999. History of renal physiology in Germany during the 19th century.*Am. J. Nephrol.* 19: 243–256.

Hoff, H. H., L. Guillemin, and R. Guillemin. 1967. *The Cahier Rouge of Claude Bernard*. New York: Schenkmann.

Holmes, F. 1974. *Claude Bernard and Animal Chemistry.* Cambridge, MA: Harvard University Press.

Hoppe-Seyler, F. 1864. Uber die chemischen und optischen Eigenschaften des Blutfarbstoffes. *Virchows Arch. Path. Anat.* 29: 233–240.

Hunter, J. 1794. *A treatise on the Blood, Inflammation, and Gun-shot Wounds.* London: Richardson.

Huxley, T. 1881. *Lessons in Elementary Physiology.* London: MacMillan.

Huxley, T. H. 1900. *Man's Place in Nature and Other Anthropological Essays.* 1863 ed. New York: Appleton.

James, J. 1993. *The Music of the Spheres: Music, Science and the Natural Order of the Universe* New York: Springer-Verlag.

Jastrow, J. 1935. *Wish and Wisdom, Episodes and Vagaries of Belief.* New York: Appleton-Century.

Jourdain, R. 1997. *Music, the Brain and Ectasy; How Music Captures our Imagination.* New York: William Monrow.

Jowett, B. 1953. *The Dialogues of Plato.* Oxford: Oxford University Press.

Kardel, T. 1997. Four episodes and a dialogue between Stenson and Borelli on two chief muscular syatems. *Acta Anat.* 159: 61–70.

Katz, L. N. and H. K. Hellerstein. 1982. Electrocardiography. Pages 265–354 in A. P. Fishman and D. W. Richards eds., *Circulation of the Blood: Men and Ideas.* Bethesda, MD: American Physiological Society.

Keynes, G. 1966. *The Life of William Harvey.* Oxford: Claredon Press.

Koenig, R. 2000. Reopening the darkest chapter in German science. *Science* 288: 1576–1577.

Kohler, R. 1973. The enzyme theory and the origin of biochemistry. *ISIS* 181–189.

Kottler, M. J. 1974. Alfred Russel Wallace, the origin of man, and spiritualism. *ISIS* 29: 144–192.

Kucharski, A. 1984. History of the frontal lobotomy in the United States 1935–1955. *Neurosurgery* 14: 765–772.

Larner, J. 1997. The discovery of glycogen and glycogen today. Pages 135–162. In F. Grande and M. B. Visscher eds., *Claude Bernard and experimental medicine* Cambridge, MA: Schenkmann.

Lasky, I. 1983. John Hunter: the Shakespeare of medicine. *Surg. Gynecol. Obstet.* 156: 511–518.

Lehman, C. 1859. *Lehrbuch der physiologischen Chemie,* 2nd ed. Leipzig: Engelmann Verlag.

Lewis, O. 1994. Stephen Hales and the measurement of blood pressure. *J. Hum. Hyper.* 8: 865–871.

Lifton, R. 1986. *The Nazi Doctors: Medical Killing and the Psychology of Genocide.* New York: Basic Books.

Lloyd, G. 1970. *Early Greek Science: Thales to Aristotle.* New York: Norton

Longrigg, J. 1988. Anatomy in Alexandria in the Third Century B. C. *BJHS,* 21: 455–488.

Ludwig, C. 1852. *Lehrbuch der Physiologie des Menschen.* Heidelburg, Germany: Winter.

Ludwig, C. 1852. *Lehrbuch der Physiologie des Menschen.* Leipzig, Germany: Winter.

Lusk, G. 1933. *Nutrition* New York; Hoeber.

Macallum, A. B. 1926. The paeochemistry of the body fluids and tissues. *Physiol. Rev.* 6: 316–326.

Magner, L. 1979. *A History of the Life Sciences.* New York: Marcel Dekker.

Manzoni, T. 1998. The cerebral ventricles, the animal spirits and the dawn of brain localization of function. *Arch. Ital. Biol.* 136: 103–152.

Marandola, P., S. Musitelli, H. Jallous, A. Speroni, and T. Bastani, d. 1994. The Aristotelian kidney. *Am. J. Nephrol.* 14: 302–306.

Marey, C. E. J. 1874. *Animal Mechanics: A Treatise on Terrestrial and Aerial Locomotion.*

Marshall, L. H. and H. W. Magoun. 1998. *Discoveries in the Human Brain: Neuroscience Prehistory, Brain Structure and Function.* Totowa, NJ: Humana.

Marx, K. 1906. *Capital,* 4th ed. New York: Modern Library.

Mauro, A. 1969. The role of the voltaic pile in the Galvani-Volta controversy concerning animal vs. metallic electricity. *J. Hist. Med. Allied Sci.* 24: 140–150.

Mazumdar, P. M. H. (1974). Johannes Müller on the blood, the Lymph, and the Chyle. *ISIS* 66: 242–253.

McEvoy, J. G. 1987. Causes and laws, powers and principles: the metaphysical foundation of Priestly's concept of phlogiston. Pages 55–71. R. G. W. Anderson and C. Lawrence, eds., in *Science, Medicine and Dissent: Joseph Priestly.* London: Welcome Trust.

McKendrick, J. 1889. *A Textbook of Physiology: Special Physiology.* Glasgow: Maclehose and Sons.

Medawar, P. B. and J. S. Medawar. 1983. *Aristotle to Zoos: A Philosophical Dictionary of Biology.* Cambridge, MA: Harvard University Press.

Mendelsohn, E. 1964. *Heat and Life: The Development of the Theory of Animal Heat.* Cambridge, MA: Harvard University Press.

Mezzogiorno, V., A. Mezzogiorno, and C. Passiatore. 1993. A contribution to the history of renal structure knowledge (from Galen to Malphigi). *Ann. Anat.* 175: 395–401.

Mowry, B. 1985. From Galen's theory to William Harvey's theory: a case study in the rationality of scientific theory change. *Stud. Hist. Phil. Sci.* 16: 49–82.

Mueller, M. and C. Fitch 1994. Kisii trepination: an ancient surgical procedure in modern day Kenya. *Explorers J.* 72: 10–14.

Müller, J. 1840. *Handbuch der Physiologie des Menschen für Vorlesung,* Coblenz.

Müller-Hill, B. 1993. Science, truth and other values. *Q. Rev. Biol.* 68: 399–407.

Müller-Hill, B. 1998. *Murderous Science: Elimination by Scientific Selection of Jews, Gypsies and Others in Germany, 1933–1945.* Plainview, NY: Cold Spring Harbor Press.

Natochin, Y. V. and T. V. Chernigovskaya. 1997. Evolutionary physiology: history, principles. *Comp. Biochem. Physiol. A.* 118: 63–79.

Needham, J. 1959. *A History of Embryology,* 2nd ed. Cambridge, UK: Cambridge University Press.

Nichols, T. 1872. *Human Physiology: The Basis of Sanitary and Social Science.* London: Nichols.

Nuland, S. B. 2000. *The Mysteries Within.* New York: Simon and Schuster.

O'Boyle, C. 1992. Medicine, God and Aristotle in the early universities: prefatory prayers in late medieval medical commentaries. *Bull. Hist. Med.* 66: 185–209.

Olmstead, J. M. D. and E. H. Olmstead. 1952. *Claude Bernard and the Experimental Method in Medicine.* New York: Henry Schuman.

Osler, A. 1921. *The Evolution of Modern Medicine*. New Haven: Yale University Press.

Pachter, H. 1951. *Magic into Science; the Story of Paracelsus*. New York: Henry Schuman.

Palfreman, J. 1976. William Crookes: spiritualism and science. *Ethics Sci. Med.* 3: 211–227.

Park, L. D. 1998. *Wonders and Order of Nature*. New York: Zone.

Pernick, M. 1996. *The Black Stork*. New York: Oxford University Press.

Piccolino, M. 1997. Luigi Galvani and animal electricity: two centuries after the foundation of electrophysiology. *Trends Neurosci.* 20: 443–448.

Pinch, T. 1996. Science as Golem. *Academe* 82: 16–18.

Ploetz, A. 1895. *Grundlinien einer Rassenhygiene. Vol.1, Die Tüchtigkeit unserer Rasse und der Schutz der Schwachen*. Berlin.

Polyak, S. L. 1941. *The Retina*. Chicago: University of Chicago Press.

Proctor, R. 1988. *Racial Hygiene: Medicine Under the Nazis*. Cambridge, MA: Harvard University Press.

Provine, W. B. 1986. Genetics and race. *Am. Zool.*, 26: 857–887.

Räumer, A. 1987. Zum Verhältnis von Religion und Zoologie im 17. Jahrhundert (William Harvey, Nathaniel Highmore, Jan Swammerdam). *Ber. Wissenschaftgesch* 10: 69–81.

Rawlings, C. E. and E. Rossitch. 1994. The history of trephination in Africa with a discussion of its current status and continuing practice. *Surg. Neurol.* 41: 507–513.

Reilly, P. 1987. Involuntary sterilization in the United States: a surgical solution. *Q. Rev. Biol.* 62: 153–170.

Reitlinger, G. 1961. *The Final Solution: The Attempt to Exterminate the Jews of Europe, 1939–1945*. New York: Perpetua.

Rheinberger, H.-J. 1987. Zum Organismus der Physiologie im 19. Jahrhundert: Johannes Müller, Ernst Brücke, Claude Bernard. *Med. Hist. J.* 22: 342–352.

Richet, C. 1923. *Thirty Years of Psychical Research: Being a Treatise on Metaphysics*. New York.

Rose, F. C. 1997. The history of head injuries: an overview. *J. Hist. Neurosc.*, 6: 154–180, © Swets & Zeitlinger.

Rosenbaum, R. 1995. The great ivy league nude posture photo scandal. *The NY Times Mag.* pp. 26–35.

Rosenfeld, L. 1985. The last alchemist-the first biochemist: J. B. van Helmont. *Clin. Chem.* 31: 1755–1760.

Ross, W. D. 1942. *The Student's Oxford Aristotle*. Vol. 2. London: Oxford University Press.

Rothschuh, K. 1973. *History of Physiology*. New York: Krieger.

Rowlett, R. 1999. Fire use. *Science* 284: 741.

Russell, B. 1945. *A History of Western Philosophy*. New York: Simon and Schuster.

Sarton, G. 1952. *Ancient Science through the Golden Age of Greece*. Cambridge, MA: Harvard University Press.

Sarton, G. 1952. *A History of Science*. Cambridge, MA: Harvard University Press.

Sarton, G. 1954. *Galen of Pergamon*. Lawrence, KS: University of Kansas Press.

Sarton, G. 1959. *Hellenistic Science and Culture in the Last Three Centuries BC*. Cambridge, MA: Harvard University Press.

Schiller, F. 1997. The cerebral ventricles: from soul to sink. *Arch. Neurol.* 54: 1158–1162.

Schmidt-Nielson, K. 1991. Bohr effect: should it be Krogh effect? *NIPS* 6: 287–288.

Schubert, C. 1988. Organisches Leben als Kreismetaphorik in der Naturphilosophie F.W.J. Schellings. *Sudhoffs Archiv.* 72: 154–159.

Schubert, W. 1996. The theory of and experimentation into respiratory gas exchange: Carl Ludwig and his school. *Pflugers Arch.* 432: R111–119.

Siegel, R. E. 1968. *Galen's System of Physiology and Medicine*. Karger, Basel and New York.

Singer, C. 1917. The scientific views and visions of Saint Hildegard : in *Studies in the History and Method of Science*. Oxford, UK: pp 1–58.

Singer, C. 1957. *A Short History of Anatomy from the Greeks to Harvey*. 2nd ed. New York: Dover.

Singer, C. 1958. *From Magic to Science: Essays on the Scientific Twilight*. New York: Dover.

Singer, C. 1959. *A Short History of Scientific Ideas to 1900*. 2nd ed. London: Oxford University Press.

Singer, P. 1979. *Practical Ethics*. Cambridge, UK: Cambridge University Press.

Smith, H. 1982. Renal physiology: in (A. P. Fishman and D. W. Richards, eds). *Circulation of the blood: Men and Ideas*. Bethesda, MD: American Physiological Society, pp. 545–606.

Sokal, A. 1996. Transgressing the boundaries: Towards a transformative hermeneutics of quantum theory. *Social Text* 46/47: 336–361.

Sokal, A. and J. Brichmont. 1998. *Fashionable nonsense: Postmodern Intellectual's Abuse of Science*. New York: Picador.

Stevens, J. 1996. Gynaecology from ancient Egypt: the papyrus Kahun: a translation of the oldest treatise on gynaecology that has survived from the ancient world. *J. R. Soc. Med.* 89: 467–473.

Stiefel, T. 1977. The heresy of science: a twelfth-century conceptual revolution. *ISIS* 68: 347–362.

Stinton, D. 1968. *Scientists and Amateurs: a History of the Royal Society*. New York: Greenwood Press.

Sullivan, R. 1996a. Thales to Galen: a brief journey through the rational medical philosophy in ancient Greece. Part 1: Pre-Hippocratic medicine. *Proc. R. Coll. Physicians Edinb.* 26: 135–142.

Sullivan, R. 1996b. Thales to Galen: a brief journey through the rational medical philosophy in ancient Greece. Part II: Hippocratic medicine. *Proc. R. Coll. Physicians Edinb.* 26: 309–315.

Sullivan, R. 1996c. Thales to Galen: a brief journey through the rational medical philosophy in ancient Greece. Part III: Galenic Medicine. *Proc. R. Coll. Physicians Edinb.* 26: 487–499.

Sutton, G. 1981. Electric medicine and Mesmerism. *ISIS* 72: 375–392.

Swainson, W. 1838. *The Natural History of fishes, Amphibians, and Reptiles or Monocardian Animals*. London: Longman and Taylor.

Tenney, S. 1996. The lives and works of two contributors to medieval physiology. *News Physiol. Sci.* 11: 292–298.

Throop, P. 1998. *Hildegard von Bingen's Physica*. Rochestor, VT: Healing Arts.

Tranoy, J. 1996. Lobotomy in Scandinavian psychiatry. *J. Mind. Behav.* 17: 1–20.

Underwood, E. A. 1972. Franciscus Sylvius and his Iatrochemical school. *Endeavour* 31: 73–76.

Vertosick, F. T. 1997. Lobotomy's back. *Discover* (August).

Vogel, S. 1992. *Vital Circuits*. New York: Oxford University Press.

Wallace, A. R. 1914. *The World of Life: A Manifestation of Creative Power, Directive Mind and Ultimate Purpose*. London: Chapman and Hall.

Wear, A. 1977. The spleen in renaissance anatomy. *Med. Hist.* 21: 43–60.

West, J. 1984. Stephen Hales: neglected respiratory physiologist. *J. Appl. Physiol.* 57: 635–639.

White. 1954. *The Book of the Beasts: Being a translation from a Latin Beastiary of the Twelfth Century*. New York: Putman.

Whitteridge, G. 1971. *William Harvey and the Circulation of the Blood*. London: MacDonald.

Willis, R. 1847. *The Works of William Harvey*. London: Sydenham Soc.

Wilson, C. 1995. *The Invisible World. Early Modern Philosophy and the Invention of the Microscope*. Princeton: Princeton University Press.

Wiltse, L. L. and T. G. Pait. 1998. Historical perspective. Herophilus of Alexandria (325–255 BC.) The father of anatomy. *Spine* 23: 1904–1914.

Wolpert, L. 2000. The well-spring. About 3000 years ago the Greeks invented science. *Nature* 405: 887.

Zinsser, H. 1934. *Rats, Lice and History*. Boston: Little Brown.

List of Illustrations

Fig. 1.1. Graphic descriptions of nature are the most ancient human intellectual activity known. Illustrations from the Chauvet caves in Southern France, which contain the oldest known cave paintings, about 30,000 years old (Chauvet et al., 1996). Reproduced with permission of the French Ministry of Culture and Communication, Regional District for Cultural Affairs-Rhône-Alpes Region, Regional Department of Archeology. **(A)** Fine renderings of heads of the long-extinct auroch. **(B)** A herd of rhinoceroses. The decreasing size of the horns and the multiplication of the lines depicting the backs suggest perspective. **(C)** Stencil of the right hand of a 30,000-year-old individual reveal a concept of self. The outline was made by pulverizing pigment on the hand flattened against the wall. (*See* color plates appearing in the insert following p. 82.)

Fig. 1.2. Boring a hole in the skull, trephination, appears to be an attempt to treat (mental) sickness, and is by far the most ancient medical surgery known. Prehistoric methods of skull trephiny **(1)** scraping, **(2)** grooving, **(3)** boring, **(4)** rectangular incisions (From Rose, 1997). Reproduced with permission by © Swets and Zeitlinger.

Fig. 1.3. Trephiny was still a major remedy in the sixteenth century. Heroic trephiny in the Renaissance (1528).

Fig. 1.4. Trephination is still practiced today in parts of Africa, e.g., by the Kisii tribe in Kenya. (Reproduced with permission from Mueller and Fitch, 1994). **(A)** A cross-shaped incision through the skin is made with a razor, the scalp flaps are reflected, and a hole is scraped through the skull bone with a hack saw. **(B)** After the surgery, the skin flaps are replaced and smeared with petroleum jelly. (*See* color plates appearing in the insert following p. 82.)

Fig. 2.1. Religious explanations of natural phenomena such as thunder are universal. Wooden sculpture of the Nigerian Yoruba tribe god of thunder, Shango, holding his spear. (From Peter Lutz, 2001.)

Fig. 2.2. By as early as 3000 BC, hepatoscopy had become an extraordinarily complex art in Babylonian prognostication. A clay model of a sheep's liver used to teach divination in ancient Babylonia, about 2000 BC. Anatomical features are labeled by Singer (1957). Reproduced with permission from Dover Publications.

Fig. 2.3. Imhotep, who lived in the twenty-seventh century BC, was later worshipped as the god of medicine in Egypt, and identified with the Greek god of medicine, Asclepius. A representation of Imhotep, the Egyptian god of medicine, Temple of Ptah at Karnak. (From Hurry, 1928.)

Fig. 2.4. The earliest representation of circumcision. From the Necropolis of Sakkara, dated at the beginning of sixth dynasty. (From Castiglioni, 1958.) A stone knife was used for the operation.

Fig. 2.5. Hieroglyphic of case six in the Edwin Smith surgical papyrus. (From Breasted, 1930.)

Fig. 2.6. Some early Egyptian animal sculptures show a beautiful naturalism. A 3500 year old bas-relief, from the temple of Assassif, of an Egyptian panther.

Fig. 3.1. Thales' conception of the floating earth. He envisioned Earth as a flat disk floating on water, with a bowl or firmament of water above our heads.

Fig. 3.2. To Pythagoras, music was number (proportion), the universe was number, so the universe was music. The concept is illustrated in Thomas Fludd's (1547–1637) *Musical Harmony of the Universe* (from *Monochordium mundi* [1623]). A Pythagorean monochord is divided into the basic harmonic intervals, each of which is an element of the universe. The earth is low G.

Fig. 3.3. A fifth century BC illustration of Achilles binding the wound of Patroclus. (From the National Library of Medicine.)

Fig. 3.4. A diagrammatic representation of the Hippocratic balance of the four humors and elements (based on Singer, 1957). The Empedoclean four-element system was enriched by the addition of four biological humors: blood, yellow bile, black bile, and mucus (phlegm). These four humors relate directly to the four elements, and the humors were responsible for the four basic human temperaments: sanguine, choleric, melancholic, and phlegmatic.

Fig. 4.1. Plato's theory of vision. Visual rays are emitted from the eye and interact with external light to form a cone of vision. When this cone touches an object, rays are reflected back to the eye. (From Grüsser and Hagner, 1990.)

189

Fig. 4.2. Aristotle's classification of animals according to the possession of (red) blood and embryological development. (From Needham, 1950.) Reproduced with permission from Cambridge University Press.

Fig. 4.3. A representation of Aristotle's *scala natura*. (From Singer, 1957.) Nature is envisioned as an orderly ladder with progressive steps, from the inanimate to plants, from the lower animals to the higher, to man and spirits, and finally God. Man is half animal, half spirit. Reproduced with permission from Dover Publications.

Fig. 4.4. Aristotle's conception of embryogenisis. (The illustration is from Jacob Rueff's *De Conceptu et Generatione Hominus* [1554] arranged by Singer [1957].) At **(a)** the menstrual blood is gathered in the womb; **(b)** shows the hot semen, with flames coagulating the blood; **(c)** note the first appearance of the blood vessels; **(d)** the first moving principle, the heart; **(f)** the "sketch" outline of the child; and **(g)** the child sculpted in more detail. (Reproduced with permission from Dover Publications.)

Fig. 5.1. Greek illustration of human operation (or perhaps dissection). (From the National Library of Medicine.)

Fig. 6.1. Galen wished to demonstrate universal harmony in body structure by proving that "the organs are so well constructed and in such perfect relation to their functions that it is impossible to imagine anything better." Galen conducts physiological experiments on a live pig tied to a table. (From Galen Venetiis, 1586 [National Library of Medicine].)

Fig. 6.2. Model of Galen's triadic plan of physiology. A detailed commentary is in the text. (Drawn by Thompson; from Peter Lutz, 2001.)

Fig. 6.3. The filtering kidney, according to Galen. The kidney was thought to consist of two chambers, an upper and lower, separated by a membrane, the colatorium, containing minute pores. Blood was supplied to the upper cavity from the renal vessels; it was purified as lighter and noxious fluid was filtered off by the colatorium. The filtrate was collected in the lower cavity, and conducted to the bladder via the ureters. (From Vesalius's *Fabrica*, 1543.)

Fig. 6.4. Chart for the examination of urine (uroscopia). A wheel of urine flasks is depicted, each with an accompanying analysis of their contents. The physician sits in the center making his diagnosis. (From Ulrich Pinder, 1506 [National Library of Medicine].)

Fig. 7.1. Medieval scribe at his desk. (From White, 1954.)

Fig. 7.2. Illustrations from a medieval bestiary (White, 1954). **(A)** The crocodile: a creature so hard that stones would only bounce off its skin. **(B)** The hyena: a loathsome beast, robbing a coffin. **(C)** The pelican: a noble bird, first justly killing its disrespectful young, but then mercifully reviving them with its own blood.

Fig. 7.3. Saint Hildegard's first vision of the universe (translated and modified by Singer, 1958). Reproduced with permission from Dover Publications.

Fig. 7.4. Hildegard's vision of the arrival of the unformed soul (Singer, 1958). The fetus appears to be fairly well developed before it receives the soul to become human. Reprinted with permission from Dover Publications.

Fig. 7.5. Hildegard's vision of death and the departure of the formed soul. Angels to the left, demons to the right, heaven above (Singer, 1958). Reprinted with permission from Dover Publications.

Fig. 8.1. Islam conserved and advanced Greek science, lost in the West. Persian human skeleton. Dated around 1200 AD (Choulant, 1920). Original in British Library, London.

Fig. 8.2. A representation of the venous and digestive systems, from a Persian manuscript dated about 1400. (From Choulant, 1920.)

Fig. 8.3. Schematic eye showing details of eye and brain connections including the optic chiasma. From Alhazens (965–1039) book of optics. This may be a copy of a Greek original. (From Polyak, 1941.) Reproduced with permission from the University of Chicago Press.

Fig. 8.4. The famous Table of Temperaments from Avicenna's *Canon*. (Translation by Gruner, 1930.)

Fig. 8.5. Western medieval copy of Arabic original. Fourteenth century Latin illustration of human skeleton. (From Choulant, 1920.)

Fig. 9.1. Diagram of the three ventricles and their contents. The legends in the ventricles are: "sensus communis, fantasia/imaginativa, vermus, cogitativa/estimativa, memorativata." (From G. Reisch *Margarita philosophiae*, Freiburg, 1503.)

Fig. 9.2. Medieval concept of the great chain of being. Divine light illuminates both angels and man, who are linked, in a perfect circle, through the kingdoms of animals, plants, minerals, and unformed matter. (From *De Intellecto*, Charles Borillus, 1470–1550.)

Fig. 9.3. The formal anatomy lecture. The lecturer sitting on his high chair reads out loud from a text derived from Galen's anatomy. The dissector, whose dress is distinguished by a row of buttons, is about to cut open the chest. His task is to illustrate the truths of the text. This drawing is from Johannes de Ketham's *Fasciculus medicinae*, 1493. It proved very popular, and was copied often, with variations.

Fig. 9.4. The parts of the eye, from a late fourteenth century manuscript. Although it has a distinctly "surrealistic" perspective, the figure contains some accurate information. The labels on the right list the three humors and the seven tunics, including the retina (**1**) and the cornea (**6**). On the left, the cranium, dura mater, pia mater and the cerebellum are named. (Choulant, 1920.)

Fig. 9.5. A late fifteenth century dissection. The bleeding indicates that the corpse is fresh. The inevitable accompanying dog looks on. (From *De propietatibus rerum* by Bartholomeus Angelicus, Bibliothèque National, Paris.) (See color plate appearing in the insert following p. 82.)

Fig. 9.6. Visceral anatomy from Gregor Reisch's *Margarita philosophica*, 1496. The thoracic and abdominal cavities have been dissected to show the most important organs in their natural settings. (From the National Library of Medicine.)

Fig. 10.1. Leonardo's sketch of the brain ventricles (ca.1506). With characteristic ingenuity, Leonardo injected molten wax into the ventricles to discover their shape when the wax hardened.

Fig. 10.2. Leonardo's drawing of human copulation. The figure has many erroneous "Galenisms." For example, he depicts a hollow tube structure between the lower spinal cord and the penis to allow for the passage of semen made in the brain to join the ejaculate.

Fig. 10.3. The magnificent frontispiece of Vesalius's *Fabrica* (1543). Eager to reveal anatomical errors, and to discover the truth, Vesalius stands next to the female body that he is about to dissect. The abdominal cavity has been opened. People of different classes throng about, arguing. The bearded man on the right sternly disapproves.

Fig. 10.4. Vesalius's weeping skeleton. Choulant (1920) comments that it "makes one think of the mourning apostles in a *'Burial of Christ'* by Titian."

Fig. 10.5. Muscle Tabula from *Fabrica*. Muscles are often depicted in a state of contraction, and suggest movement and activity.

Fig. 10.6. Brain dissection from *Fabrica*. Note the emphasis is still on the hollow brain ventricles.

Fig. 10.7. A sixteenth century anatomical study of the outer muscle layer, from a front view of the body. This bizarre illustration of a man who has skinned himself is from *di Hamusco* (1556) by Juan Valverde (Choulant, 1920).

Fig. 10.8. There was a revival of interest (an Aristotelian interest) in discovering nature seen in such studies as a dolphin and its placenta. (From *Histoire naturelle des etranges poissons marins* [1551] by Pierre Belon.)

Fig. 10.9. Innovatively, the skeletons of a bird and man are compared bone by bone. (From Pierre Belon, *Histoire de la nature des oyseaux*, 1555.)

Fig. 10.10. Dürer's magnificently, but imaginatively, detailed rhinoceros (1515).

Fig. 10.11. A study of the dogfish. (From Guillaume Rondelet, *De piscibus marinis*, Lyons, 1554.)

Fig. 10.12. Old ideas and old creatures tenaciously held on. The dreadful lamia, with a beautiful woman's face and a serpent's tail, could entice young men to approach so that she might feed on their blood. (From Conrad Gesner's [1516–1565] *Historiae animalium*, translated and expanded by Edward Topsel [1572–1638] as *The History of Four Footed Beasts*, 1607.)

Fig. 10.13. A seventeenth-century illustration of the cerebrum, in which, still following the ancient Greeks, the convolutions are drawn to resemble intestinal coils (Caserio, 1627).

Fig. 10.14. The medieval world still rules in Fludd's world of the brain (1617). The three ventricles house sensation, intellect, and memory/motion in a complex interrelationship that links to the macrocosm, and to God and his angels.

Fig. 10.15. The revolutionary new water closet had its hazards. (From *The Metamorphosis of Ajax* by the inventor, John Harrington, 1556.) Note fish in the water reservoir.

Fig. 11.1. Frontispiece of Harvey's *De motu cordis* (1628).

Fig. 11.2. Harvey's demonstration in *De motu cordis* of the action of the venous valves, which only allows blood to flow to the heart. Tying a light tourniquet at **A** causes, "especially in laborers," certain knots or elevations to be seen in the veins (**B–F**). Pressing one of these valves with a fingertip, and with another finger pushing the blood upwards/towards the heart, you will see this part of the vein (**O, M**) stays empty, and that the blood cannot flow back.

Fig. 12.1. Seventeenth century biology in France. The inner organs of a fox are being studied. Seated at right, Claude Perrault points to a page of one of his animal monographs. To the left of the picture the secretary keeps the minutes. (From *Jardin des plantes* 1669, by LeClerc.)

Fig. 12.2. Sanctorius's balance for carefully comparing the weight of solid and liquid ingested with that excreted, and changes in body weight. (From his *De statica medicina*, Venice, 1614.)

Fig. 12.3. Sanctorius's Balneatorium: a leather bag that allows flowthrough bathing of the patient in bed, and the collection of such fluid.

Fig. 12.4. Descartes' hydraulic model of the brain causing muscle contraction. The pineal (**H**) secretes animal spirit into the ventricle (**E–E**), which then makes its way through a network of small pores (**a**) through the walls of the ventricle (**A**). The spirit passes down tubes (**B**) to the spinal cord (**D**), and hence to the muscle, which swells causing contraction. Changes in the inclination of the pineal controls the amount of animal spirit flowing to the muscle.

Fig. 12.5. Descartes' model of sensory processing. Light from the arrow enters the eyes and forms an inverted image on the retina. The image travels along the hollow nerves to the pineal gland, where it is interpreted (*Treatise on Man*, 1662).

Fig. 12.6. Descartes' mechanical-hydraulic model of sleep. (**A**) represents sleep, (**B**) awakefulness. In sleep, the flow of animal spirit from the pineal (**H**) has been reduced to a trickle, the ventricles have collapsed, and the hollow nerves are flaccid. When awake, copious amounts of animal fluid are produced, swelling the brain and making the nerves taut (*Treatise on Man*, 1662).

Fig. 12.7. Structure of muscle according to Steno's geometric model. A single muscle is viewed as a parallelogram (ABCDEFGH). The tendons at either end are prisms (DMIKLC and EFQPON). (From Bastholm, 1950.)

Fig. 12.8. Borelli's geometric models of muscle mechanics. The contracting elements, the "finest fibers" of muscle are viewed as rhombs. (From *De motu animalium*, 1685.)

Fig. 13.1. Hooke's famous flea. (From *Micrographica*, 1651.) Samuel Pepys commented in his diary, "Before I went to bed I sat up till two o'clock in my chamber reading Mr Hooke's *Microscopical Observations*, the most ingenious book I ever read in my life."

Fig. 13.2. Hooke's microscope. (From *Micrographica*, 1651.) A technological breakthrough. (National Library of Medicine.)

Fig. 13.3. Frog lungs showing capillaries (**HHH**). (From Malphigi, *De pulmonibus*, 1661.) Malphigi had discovered the connection between the arterial and venous blood systems, and so provided the missing link in Harvey's circulation theory.

Fig. 13.4. Leeuwenhoek's microscope. In this model, a biconvex lens is mounted between the metal plates. The object to be viewed is secured on the metal point.

Fig. 13.5. Leeuwenhoek's illustrations of red blood corpuscles. (**1**) Salmon; (**2–4**) eel; (**5–6**) aggregation of eel blood corpuscles. Fig. 7 (*inset*)shows the capillaries (**EF**) between an artery (**HI**) and a vein (**AB**). (National Library of Medicine.)

Fig. 13.6. Swammerdam's illustration of the amazing structure of the honey bee's compound eye.

Fig. 14.1. Hogarth's parody of anatomists in "*The Reward of Cruelty*," 1734. Cupidity, ignorance, and stupidity are well illustrated.

Fig. 14.2. Blood transfusion from a sheep to a man. (From Purmann's book on military surgery, 1721). (National Library of Medicine.)

Fig. 14.3. Caricature of the dissection room. (From Rowlandson, 1756–1827.) The main characters are actual contemporary scientists. The atmosphere is one of "unhealthy" obsession.

Fig. 14.4. Nonconformity had its price. Mob storming Priestley's house, July 14, 1791.

Fig. 15.1. Hales' data on the amount of blood that "issued on severing the carotid arteries of 25 horses," and the height it rose to in connecting tubes. (From *Haemastaticks*.)

Fig. 17.7. Sketch for a physiological experiment, from Claude Bernard's note book. (From Hoff et al., 1976.)

Fig. 17.8. Bowman's illustration of the human nephron. The Bowman's capsule (**c**) holds a tuft of capillaries.

Fig. 17.9. Abdominal hole opening directly into the stomach of Alexis St. Martin, which allowed Dr. Beaumont to perform his seminal experiments on gastric digestion, to the discomfort of Mr. St. Martin. (From Beaumont, 1833.)

Fig. 17.10. Helmholz's myograph. A short electrical stimulus was applied at (**a**). A single muscle contraction results; the second wave is an artifact.

Fig. 17.11. Du Bois' demonstration of electrical stimulus applied to the nerve transmitted to muscle (1848). The frog is tied, with nonconducting silk, on a wooden platform. Electrical current is applied to the spinal cord by means of two brass clamps.

Fig. 17.12. Amputation of the leg without anesthetic. (From Gersdorf, Strassbourg, 1517.)

Fig. 17.13. Administering chloroform. (From Snow, 1858.)

Fig. 17.14. Marey's apparatus for recording head and foot movements in a running man. (From Marey, 1874.)

Fig. 17.15. Beginnings of high altitude physiology. The hydrogen balloon allowed a rapid ascent to high altitudes. On 5 September 1862, James Glaisher (at right), chief meteorologist at the Royal Observatory at Greenwich, recorded the onset of physical and mental impairment brought about by the increasingly severe hypoxia, until he collapsed unconscious at an altitude of seven miles. Some of the accompanying pigeons appear unperturbed.

Fig. 18.1. Louis Agassiz's "physiological" interpretation of God's plan on the unfolding of life on earth for the preparation of man. (From Agassiz and Gould, 1851.) (*See* color plate appearing in the insert following p. 82.)

Fig. 18.2. The Quinarian system of natural classification, based on circular analogies. (From Swainson, 1838.)

Fig. 19.1. Ectoplasm, Nobel Prize winner Professor Richet's newly discovered life form that, in contrast to endoplasm, materializes outside the body, and can assume human personification. (From Jastrow, 1935.)

Fig. 19.2. Professor Crooke's apparatus for measuring psychic force. One spirit caused an 80-pound displacement. (From Crooke, 1874.)

Fig. 19.3. Crooke's proof of spirit musicality. The accordion suspended in a basket is held by one hand; tunes are played by spirits. (From Crooke, 1874.)

Fig. 19.4. Dr. Freeman demonstrating his ice-pick lobotomy technique. (From Corbis Images.)

Fig. 19.5. Eugenics Society poster warning about the dangers of broadcasting bad seed. (*See* color plate appearing in the insert following p. 82.)

Fig. 19.6. English translation of Nazi Hereditary Health Court Questionnaire. (From Lifton, 1986.)

INDEX

Biography

Peter L. Lutz, PhD holds the McGinty Eminent Scholar Chair in Biology at Florida Atlantic University. Dr. Lutz received both his BSc and PhD from Glasgow University, Scotland. He has held faculty positions at the University of Ife, Nigeria, Duke University, USA and the University of Bath, England. He was chair of Marine Biology at the Rosensteil School of Marine and Atmospheric Science, University of Miami, before taking up his present position in 1991.

As a comparative physiologist he has worked on the physiology of a wide variety of organisms, from liver flukes to duck-billed platypuses. His current interests center on the survival mechanisms of anoxia tolerant brains. He has a long and deep interest in the history of science, and has taught courses at the undergraduate and graduate levels. Dr. Lutz is a member of the Society for Neuroscience, the American Physiological Society, the Society for Experimental Biology and is a Fellow of the Explorers Club. He is a Series Editor for the *Marine Biology Series* published by CRC Press and is a Senior Editor for the *Journal for Experimental Biology.* He has authored more than 150 research papers and 3 books.